Manohar Dingari

Analytical Study of Air Traffic Using ARFIMA Time Series Models

Manohar Dingari

Analytical Study of Air Traffic Using ARFIMA Time Series Models

LAP LAMBERT Academic Publishing

Cover image: www.ingimage.com

Publisher:
LAP LAMBERT Academic Publishing
is a trademark of
Dodo Books Indian Ocean Ltd. and OmniScriptum S.R.L publishing group

120 High Road, East Finchley, London, N2 9ED, United Kingdom
Str. Armeneasca 28/1, office 1, Chisinau MD-2012, Republic of Moldova, Europe
Managing Directors: Ieva Konstantinova, Victoria Ursu
info@omniscriptum.com

Printed at: see last page
ISBN: 978-620-8-44468-6

Dedicated

To

My Family

&

All My Well Wishers

ACKNOWLEDGEMENTS

I would always grateful to acknowledge the enthusiastic and effective supervision of Dr. **D. Mallikarjuna Reddy,** Assistant Professor, Department of Mathematics, School of Science, GITAM (Deemed to be University), Hyderabad who stood behind me as a tower of strength and whose knowledge, analytical ability and I owe an ineffable debt of gratitude to him for his guidance and support throughout my research study. He has been the main driving force for encouraging me to take up this work and complete it successfully.

I am thankful to **Prof. K. Siva Rama Krishna** Vice-Chancellor, **Prof. K. V. G. D. Balaji** Registrar and **Prof. Ch. Rama Krishna**, Director, UGC Affairs, GITAM (Deemed to be University) for providing me an opportunity to fulfill my long cherished dream of pursuing Research.

I convey my deep sense of the gratitude to **Prof. N. Siva Prasad**, Pro Vice Chancellor, GITAM (Deemed to be University), Hyderabad for his excellent support and encouragement during the course work.

I would like to acknowledge the support and encouragement of **Prof. Ramarao**, Principal School of Science, Hyderabad.

I gratefully acknowledge **Prof. M. Saratchandra Babu** Principal, Institute of Science GITAM Institute of Science, GITAM (Deemed to be University), has helped and encouraged me during my long journey in pursuing a Ph. D. degree.

I express my heartiest thanks to **Prof. A.Chandra Sekhar**, HOD & **Prof. N Ravi Shankar** Department of Mathematics, GITAM Institute of Science, GITAM (Deemed to be University), for providing facilities and information to carry out the research work.

My Sincere thanks to **Prof. K. Maruthi Prasad**, Head, **Dr. Siva Reddy Sheri**, Associate Professor, and all other faculty members, Department of Mathematics , School of Science, GITAM (Deemed to be University), Hyderabad campus, for their ideas, insights and comments which helped me to find a path for this study.

I express my lifelong gratitude to my father, mentor, guide and philosopher **Shri D. Satyanarayana Chary,** who has given me life and provided me with all facilities which enabled me to reach this stage. He stood behind me as a tower of strength, and without whom this study would not have been possible. My lifelong gratitude to my mother **Smt. Padma** who has given me life and a constant source of support.

My lifelong thanks are also due to my life partner and my beloved wife **Mrs. Sumalatha Dingari,** who inspired me and supported me in all my endeavors and without whom I am not a complete person. She has encouraged and supported me in many ways in completion of this thesis. I also thank my daughter **Baby Snigdha Srikari** who has been my source of support.

I would like to thank Management of Sri Sai Degree & PG College, Secretary & Correspondent **Dr. Palla Rajeshwar Reddy**, Academic Director **Prof. M. Govinda Ram Reddy,** Principal Shri **Ch. Dilip** and Shri **T. Nagaraju** for their support, who emboldened me and inspired me until end of my thesis. I also thank all my family members, friends, relatives for their unconditional support at each turn of the road.

Finally, but all above, I offer my humble respect to almighty for blessing me with courage, strength and mental stability to achieve this endeavour. This thesis is a small tribute to all those who work for peace and harmony.

(Dingari Manohar)

CONTENTS

LIST OF TABLES

List of Figures

LEGENDS

Abbreviations	Expansion
DGCA	Directorate General of Civil Aviation
LAN	Local Area Network
WAN	Wide Area Network
WWW	World Wide Web
LRD	Long Range Dependence
SRD	Short Range Dependence
FBM	Fractional Brownian motion
RMD	Random Midpoint Displacement
MAP	Markovian Arrival Process
QoS	Quality of Service
FGN	Fractional Gaussian Noise
USA	United States of America
CPU	Central Processing Unit
GRN	Gaussian Random Numbers
ACF	Auto Correlation Function
PACF	Partial Auto Correlation Function
PSD	Power Spectral Density
HW	Holt-Winters
ANN	Artificial Neural Network
MLP	Multi-Layer Perceptron
MSE	Mean Square Error
RMSE	Root Mean Square Error
MAE	Mean Absolute Error
MAPE	Mean Absolute Percentage Error
BIC	Bayesian Information Criterion

ACRONYMS

Acronyms	Expansion
LAN	Local Area Network
WAN	Wide Area Network
ARMA	Auto Regressive Moving Average
ARIMA	Auto Regressive Integrated Moving Average
SARIMA	Seasonal Auto Regressive Integrated Moving Average
ARFIMA	Auto Regressive Fractionally Integrated Moving Average
FARIMA	Fractional Auto Regressive Integrated Moving Average
MAP	Markovian Arrival Process

ABSTRACT

Many researchers have worked on Time series forecasting techniques. However, despite its importance, the Self-similar structure has not been addressed. Many techniques are developed for forecasting. The purpose of this research is to investigate the self-similar structures in the Air Traffic, Real-time air traffic data of Air India and Indigo Scheduled Domestic flights and develop a suitable forecasting model to Self Similar Time series data. Also, the concept of self-similarity has now proved itself to be extraordinarily useful. This research describes the current state of knowledge about the self-similar process, and their applications, concept, definitions and basic properties are emphasized. Self-similarity covers alongside statistical estimation. ARFIMA, Long Range Dependence, Hurst parameter are driven by Self-similar processes. It will find an ideal entry point for studying the already extensive theory and applications of Self-similarity, giving the researchers a road map of the realm of self-similarity that allows for further exploration.

This thesis is a collection of chapter contributions that bring together relevant past works spanning a cross-section of topics covering traffic modelling by Box- Jenkins methods, Exponential Smoothing methods and Artificial Neural Networks for air traffic. The thesis's primary objective is to present a comprehensive yet cohesive account of some of the principal developments and results concerning time series forecasting techniques to serve as a reflective milestone that captures the state-of-the-art field. The thesis is organized around three main subtopics—air traffic modelling by existing time series forecasting models, investigating Long Range Dependence of air traffic through Self Similarity and fitting ARFIMA(FARIMA) to the Long Range Dependent time series then compare with existing models and identify the best model for air traffic.

CHAPTER-I

1.1. INTRODUCTION

Air transport is presently a quickly developing service sector in India. People continuously travel for various reasons (tourism, business) to different countries and worldwide. The rapid growth of demand for air transport for the past few decades and the deregulation policy has affected other air carriers in increasing transport services; thereby, airports and airways face huge congestion problems. As a result, analyses and forecasts of data regarding air passengers are of great importance in designing development strategies for air service providers. Air transport improves the growth of the local economy, arriving in the development of companies, and thus developing the competitiveness of a business. Therefore, it is crucial to analyze the future air traffic and its changes with time properly, which helps the air service providers design their services adequately. Forecasting traffic intensity using air traffic models involves finding the future demand for air transport in the service. Another important thing is the seasonality of air traffic. Seasonal variations make it essential to observe traffic intensity every month of the year and on each day of a month. The immediate result of a traffic forecast includes the number of air passengers. Hence the significance region that we suppose throughout the thesis is preparing a tool to forecast the air traffic.

Analysis of time series has become a progressive research field for the past few decades, which attracted most of the researchers' attention. The principal focus in developing a time series model is collecting data cautiously and thoroughly investigates the pattern of previous data of the series to establish a proper model that reports the essential pattern of the series. Then it can be applied to forecast the future pattern. Thus time series forecasting may be treated as understanding the past and predicting the future because of this essential significance of forecasting using time series in various spheres like science, engineering, finance, economics and business and so on. Extreme attention has to be given to developing a suitable time series model for the data under consideration. A proper fitting of the model leads to adequate forecasting of the time series. Researchers have put vigorous attempts in developing effective models which increase the precision of forecasting.

Consequently, many models for time series are included in linear and non linear models for forecasting future demand in the literature. All these models discuss similar fundamental idea, but they accompany various patterns. The fore studies reveal that any forecasting model may not be best for every case for all the conditions. However, fore studies discussed various studies for comparison, and they do not propose one particular method as superior to another. Accordingly, seasonality, difficulty, and trend present in the air traffic require analysis about the most suitable method for each phenomenon (Box and Jenkins, 1970; Hipel et al., 1994).

Auto Regressive Integrated Moving Average (ARIMA) is a very famous and widely applied stochastic model for time series. Execution of ARIMA assumes that the data under consideration is linear, also has the characteristics of Normal distribution. Subclasses of the ARIMA model are Autoregressive Moving Average (ARMA), Autoregressive (AR) and Moving Average (MA) models. For forecasting the time series with seasonality, Box-Jenkins had given SARIMA (Seasonal Autoregressive Integrated Moving Average). ARIMA models have become popular because of their adaptability to describe different sorts of data with clarity and optimal model building processes with Box-Jenkins methodology.

Another most widely used forecasting method is the Exponential smoothing method which is extensively used in business and industry. Especially the Holt-Winters methods analyze time series with one variable that studies seasonal factors and trend. These methods are simple to formulate, and they became more popular. Forecasting comparisons have proved that the forecasts of Holt-Winters methods are more accurate, obtained with simple model formulation and computation. The results are either similar or better than the results obtained by ARIMA models. The smoothing parameters of the method are generally calculated by minimizing the mean square errors one step ahead after fixing the initial values.

In recent years, in the sphere of forecasting the time series, Artificial Neural Network (ANN) has won many researchers attention. Because of its capability of representing any non linear or linear function and do not require any assumption of a probability distribution of the time series (Santos *et al.* 2014, Kunt *et al.* 2011 and Claveria, Torra

2014). Initially, ANNs were inspired by the biological neuron system, but recently ANNs have successfully applied in many areas, particularly for forecasting and classification purposes. This approach uses a lot of elementary processing units, depending on the pattern of the human brain, it learns the real-time series by training a computer (Haykin 1999). For the last few years, a lot of research has been done to develop a forecasting model by applying neural networks. In the literature, different forecasting approaches of ANN are available. The Multi-Layer Perceptron (MLP) is the most widely used approach among them. It is described by a single hidden layer Feed Forwarded Network (FFN). As a result of continuous ongoing research works, a lot of neural network patterns are existing. The present thesis mainly focuses on the ANN as mentioned above forecasting model.

Statistical analysis of traffic measurements from networks like Ethernet LAN, WWW and Wide Area Network traffic have characteristics about self-similarity. The self-similarity introduced (Mandolbrot, 1982) and used in Geology and Hydrology. The Self-similar behaviour causes strict degradation in network performance and makes the traffic network is bursty (Erramalli et al., 1996). Hence there must be a time series forecasting model which can take self-similarity into account. Air transport is currently one of the fastest growing service sectors. It is of great importance to analyse and forecast air traffic in designing development strategies for air carriers. Presently the issue is to check whether self similarity is present in air traffic or not to analyse and develop a model of air traffic. Previously the road traffic has been modelled using the Poisson approach. The same has been practiced in WWW, WAN, LAN and Ethernet traffic. In general, most of the times nature of the traffic is bursty. Mathematically traffic's burstiness is studied as long-range dependence (LRD) or self similarity. The characteristics of self similarity cannot be studied by the Poisson process.

The research philosophy, in general, in this domain, is depicted in Figure 1.1. Advancement in this research area is a variant in the time series analysis.

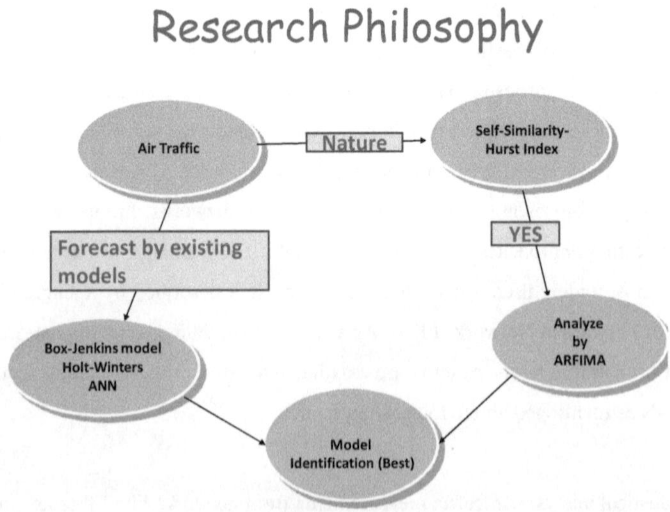

Figure 1.1 Research Philosophy

The expansion of measuring techniques on air traffic makes it feasible to analyze many traffic flows. The outcome reveals that authentic air traffic data is self-similar, signifying that the traffic perceived by different time intents has similar statistical properties. The air traffic in self-similarity is a tough disparity to conventional techniques of air traffic and point up inconsistency on short-time scales but is basically flexible on large-time scales also, they lack LRD. Because self-similar traffic has strong implications for modelling, scheming, and designing the services, there is an urgent need to find appropriate traffic self-similarity measurement parameters. Also, the reason behind the self-similarity in air traffic must be clearly acknowledged since understanding air traffic's basic features provides an opportunity for a proper design of development strategies and enhanced performance. The thesis investigates the mechanisms that persuade self-similarity in air traffic; the nature that causes of self-similarity of air traffic must be notorious because discriminating air traffic character is significant with the design implementation of air service providers.

Traffic congestion occurs wherever demands exceed the capacity of the system. The most important issue in the study of any traffic system is to predict the future requirement. ARIMA technique analyzes the short memory process. Long memory processes are analyzed by the Fractional Autoregressive Integrated Moving Average

(FARIMA) model, which have stabilized mean and variance that is stationary. ACFs of the process with long memory deteriorate very steadily than the process with a short memory. This long Range Dependence (LRD) can be best analyzed by ARFIMA (or FARIMA) models, and the principle of parsimony also satisfies.

The ARFIMA models are the generalization of ARIMA models. This research's chief intention is to prepare a tool to analyze and forecast the traffic flow month-wise, which helps air carriers revise their services.

1.2 REVIEW OF THE LITERATURE

Analysis of time series consists of the following procedure, initially verifying the data's stationarity by analyzing the autocorrelation function (ACF), identifying an appropriate model and estimating the parameters and finally forecasting the future by applying the identified model (Brockwell and Davis, 2002). This part writes-up a general analysis of the time series technique and focuses on extending the long-memory processes. ARIMA is the famous Box & Jenkins model (Box et al. 2013), and it belongs to the family of linear models that describes the stationary and non-stationary time series. Its limitation is, it describes only short range dependent time series. If a short range dependent time series is stationary, its autocorrelation function would quickly decay so quickly that its sum converges. Non-stationary time series can be converted to stationary by differencing the original time series for the required number of times. After proper differencing, the ACF should deteriorate exponentially fast to zero. However, the degree of difference can only take integers in the traditional ARIMA model.

Methods of Exponential smoothing are very famous forecasting methods applied in business and industry, especially Winters' method deals with time series with one variable that has seasonality and trend factors. They got important because they are simple to formulate, and they give accurate forecasting results (Gardner, 1985). Investigations showed that Holt and Winters methods take less effort in developing the model and calculation. The accuracy of the results was either the same or efficient as those obtained by ARIMA models (Makridakis et al., 1998).

Recently, the application of artificial neural networks has increased rapidly as they are capable of relating non linear or linear function. This approach has now applied in various disciplines, such as banking (Venkatesh et al. 2014), transportation (Jimenez et al. 2014), touristy requirement forecasting (Palmer et al. 2006, Claveria, Torra 2014), energy demand forecasting (Jarimillo-Moran et al. 2013, An et al. 2014), economics (Choudhary, Haider 2012) and prediction of accidents in traffic (Kunt et al.2011, Akgungor, Doğan 2009).

In addition to time series analysis and forecasting, there is also research related to fractals and self-similar processes, which involves estimating the self-similar parameter *H*, fractional Brownian motion and its application, and fractals in human physiology. Fractal and self-similarity have been hot topics ever since they were first introduced. They provide a fresh path to investigate the world. This chapter provides a literature review on time series analysis, especially the long-memory analysis of time series and other research related to fractal and self-similarity.

Statistical analysis of multimedia data traffic in high-speed networks, for instance, packet flows in WWW, WAN, LAN is carried out and shown to be self-similar (Leland et al., 1994). It is shown that the Poisson process fails as accurate models for WAN traffic (Vern Paxson and Sally Floyd, 1998). Superposition of Pareto distributed ON/OFF periods is proposed as a first way to generate self-similar traffic (Crovella and Bestavros, 1997; Sherman et al., 1997). The resulting aggregate traffic, in this case, displays asymptotic self-similar properties as proved in (Vern Paxson and Sally Floyd, 1998). Among the other generic self-similar techniques based on the probability distribution, FBM established a lot of concentration (Norros, 1995; Norros, 1994). The increment process of FBM is known to be FGN and was pioneered (Mandelbrot and Van Ness, 1968). However, these models fail to confine the short-range correlation of real traffic.

An approximate algorithm is based on normal distribution to generate self-similar traffic and is known as the RMD method (Lau et al., 1995). The main disadvantage of Random Midpoint Displacement is to makes the self-similar structure approximately.

There are various methods for estimating Hurst exponent, which provides the measurement of self-similarity viz;

(i) Method of Rescaled adjusted range statistics (Hurst, 1978).
(ii) Variance time method (Cox and Smith,1953)
(iii) Periodogram method (Daniell, 1948)
(iv) Correlogram method (Licklider, 1951)
(v) Higuchi's method (Higuchi, 1988) these methods are described in chapter II.

We have proposed a novel method (MallaReddy Perati et al., 2012). based on percentiles to compute the Hurst parameter; the same is applied to air traffic and is validated with two methods viz., (i) Correlogram method (ii) Variance Time method.

The ARIMA models cannot describe the persistency; therefore, they cannot be applied to the long-memory time series. For this situation, J. R. M. Hosking, in 1983, proposed the FARIMA (ARFIMA) model, which is short for fractional autoregressive integrated moving average(Hosking, 1981), an extension of the ARIMA. The FARIMA model successfully describes the long-range dependence of a time series and has been applied to model the LRD in economics and hydrology (Doukhan et al., 2003) and internet traffic data (Liu et al., 1999). In the FARIMA model, fractional differencing is introduced, which allows the degree of differencing to take fractional numbers.

(Katarzyna chudy-laskowska and Tomasz pisula, 2017) investigated forecast model to air passenger transport in an airport in south western Poland. They considered data in monthly cycles and applied four research methods to forecast- Seasonal ARIMA (SARIMA), Support Vector Machines, ANN and Seasonal Exponential Smoothing.

(Alberto Andreoni and Maria Nadia Postorino, 2006) have proposed ARIMA techniques for univariate as well as multivariate to find the air transport requirement of an airport in South of Italy named Reggio Calabria regional airport. They have studied the influence of the latest adjustments in the availability, which increases the air transport demand. Also highlighted the variations of one variable and multivariate model.

(Claudimarpereira Da Veiga et al., 2014) have compared the efficiency of Method of Winters and ARIMA for forecasting future values corresponding to a class of food items that are easily spoilt. They used U theil and MAPE as evaluation measures. By these performance measures, they showed the Holt-Winters model as better than ARIMA.

(Likhanov et al., 1995) investigated an analytical study based on ATM buffer determined with the self-similarity traffic. The probability of buffer tenancy is found, and it is shown that this probability diminishes with the size of a buffer, not exponentially, while in the conventionally Markovian traffic representations, but mathematically.

(Chitra et al., 2015) have attempted to fit the ARIMA model to supply chain management of fresh vegetables. They used the data of monthly supply and price of vegetables to find out the Seasonal factors. Based on these Seasonal patterns, they suggested developing cold storage capacities in needy places.

(Tsybakov et al., 1998) have considered the discrete-time processes second-order self-similar with several common positions of self-similar characteristics. A representation used for ATM cell data traffic be also measured the self-similarity nature setting of this technique is established.

(Etebong P Clement, 2014) has used chemical viscosity reading series to apply the time series technique. He compared the Normalized Bayesian Information Criterion with the actual Box and Jenkins model. He also presented analysis and assessment, and then proved that the proposed approach is better than the actual Box and Jenkins method.

(Vasantha Kumar and Lelitha Vanajakshi, 2015) have used restricted data to predict short term traffic flow using SARIMA. They considered a 3-lane arterial roadway in Chennai, India, selected as a study area and took limited flow data from three successive days to develop the SARIMA model. Model parameters were estimated using the Maximum Likelihood (ML) method. They compared the proposed model with the navy method and the historic average. They suggested SARIMA as the best method for the data using MAPE.

(Qin Yu et al., 2005) have contributed the three statistical analysis methods are, Periodogram analysis, R/S statistic and variance-time plots are applied to Hurst index for estimation of self-similar (LAWN) wireless traffic of LAN. The statistical attributes of group multiple self-similar input data sources at the LAWN entry point are deliberated.

(Chinwen Cheong, 2010) investigated, fractionally integrated methodology in the global financial equity market of self-similar process under the heavy-tailed supposition of the asymmetric and symmetric fractionally integrated time changeable unpredictability is developed by choosing the most suitable power transformation of the volatility proxy.

(Irina Klevecka, 2011) investigated an advanced algorithm to forecast short term traffic of electronic communications by using non-linear neural networks. They discussed time-lagged feed-forward networks for time series modeling and forecasting, using Multi-Layer Perceptron (MLP).

(Bharati Sharma et al., 2018) attempted to design an approach for forecasting short term traffic using back propagation ANN by considering two-lane undivided highway with mixed traffic conditions in India. They compared their results with support vector machine, random forest, regression tree, k-nearest neighbour classifier and multiple regression methods. They proved that back propagation Neural Networks give better performance than other techniques.

(Emrah Onder and Sultan Kuzu, 2014) focused on estimating the air traffic volume by applying smoothing techniques, also investigated seasonal effects of the parameters. They applied traditional forecasting techniques of time series like liner exponential smoothing of single and double Brown's method, moving averages, decomposition methods and Holts double exponential smoothing technique for forecasting the size of air traffic. They focused mainly on applications of traditional time series analysis.

(Adas, 1997) examined and survey the traffic models that are presently used in the text. Non-traditional Long-Range Dependent traffic and Traditional short-range models are offered. The number of parameters required, systematic tractability, parameter evaluation, and traffic representation capability to confine Marginal Distribution, auto-correlation structure for the real traffic are worked out.

(Hurst, 1951) he investigated the history of the water board in productive Crescent's closely connected to the creed. This is almost clear in ancient Egyptian astronomical time. Composed hydrological and excessive data and utilized it to excise tolls and forecast the floods that hose down the arable land. The Nile River's unique hydrological features make it rather predictable in actions compared to other major rivers of the province.

(Jose et al., 2005) used UK air passengers month-wise to fit Winters additive method for forecasting the air passengers, compared the resulted forecast with results obtained previously. They discussed the Holt-Winters additive as inappropriate when the level affects the data's seasonal components, but this method can be helpful when transforming the series adequately.

(Ghosh et al., 2004) used three various time series techniques Seasonal ARIMA model, Holt-Winters methods and random walk model to forecast the traffic flow in a congested urban transportation network in Dublin. Using these methods, they have done simulation and short term forecasting of univariate traffic flow. They established that the Holt-Winters method and Seasonal ARIMA give extremely competitive forecast and considerably similar to the actual traffic flow during peak hours.

(Mitko Gospodinov and Evgeniya Gospodinova, 2005) depend on the fractional Gaussian Noise (FGN) method, and they studied graphical methods designed for measuring Hurst exponent of pseudo-random sequences of self-similar nature. The analysis showed that the Fractional Gaussian Noise process always produces self-similar characteristics. With the virtual imprecision of the measuring Hurst exponent less than 8%, the generator should be suggested a realistic simulation study for high-speed telecommunication networks.

(Ette Harison Etuk and Tariqmahgoub Mohamed, 2014) fitted SARIMA model to rainfall data. They fitted three multiplicative SARIMA models using the autocorrelation structure and identified the most adequate model using residuals correlation.

(Qiang Meng and Hooi Ling Khoo, 2009) they have investigated the quantitative individuality of vehicle influx pattern on highways. This study shows that if the vehicles' arrival pattern on highways shows the self-similar nature, the ensuing time development distribution is carried out. Through employ real highway road traffic data had offered from the Texas Department of Transportation, the USA, this investigation examines the continued survival of self-similar characteristics lying about these arrival data of vehicles. It is ended by calculating the Hurst index, which is the measurement of self-similarity testing. The hypothesis examination from the Hurst index evaluation displays that vehicles' arrival pattern on the highway via restrained to congestion traffic conditions demonstrates the self-similarity characteristics.

(Helge Brugner, 2017) investigated the impact of varying parameters for aggregation of network traffic flow predictions using the Holt-Winters filtering algorithm. They considered changes in aggregation level and aggregation function and a change in training set length. They concluded that Holt-Winters combining with small aggregation levels and aggregation function as 90th percentile, perform well for short term prediction.

(Sarkar, 2007) Stochastic practices are fixed in distribution by prudent scaling of space and time, known as self-similarity property, representing the long-run properties of variable presenting the LRD stingily. Heuristic assessment processes have been offered here and revise subsequently that self-similarity exponent H provides the relationship formation of LRD processes can be proficiently approximated.

(Fei Xue et al., 1999) provided a methodology to develop a FARIMA technique for original congestion, also a method to produce a FARIMA process with given parameters. They fitted FARIMA models to four measured traffics and accompanied that FARIMA technique as the best approach for traffic, which can describe the behaviour of LRD and short-range dependent on real congestion. They gave some guidelines to minimize the complexity in developing the FARIMA model, which reduces the computational time.

(Penarat srisaeng et al., 2015) focused on estimating the demand and revenue passenger kilometres performed of Australia's low-cost flight using ANN and traditional multiple linear regression models. The back propagated feed forwarded MLP architecture ANN was used and the activation functions applied were linear function and sigmoid function. They declared that the ANN technique is superior to the classical regression models using the R square value.

(Mohie El-Din, 2017) employed back propagation Neural Networks and genetic algorithm to forecast the demand for International and Domestic air passengers in Egypt. They studied the factors which influence air passengers. By MSE, they concluded that the results obtained by the genetic algorithm are more accurate than back propagation neural networks.

(Bilnova, 2007) investigated the ANNs approach for forecasting the spread of the Russian network of air carriers. 28- time-lagged feed forwarded ANN technique was used for forecasting the flow of passenger. They used this model to forecast the short term network of air carriers in Russia.

(Lawrence R Weatherford et al., 2002) investigated a forecasting technique using ANN to apply in the airline industry compared with the traditional forecasting techniques like regression, exponential smoothing and Moving averages etc., by using MAPE. They proved that Neural Networks architectures provide better forecasting results than traditional forecasting methods.

(Perati et al., 2012) until recent times, for modelling road and internet traffic queues, the Poisson process has been used. These are studies through different methods; the real-time data of road traffic is shown as self-similar. And the outcome of this assessment proved that the observed traffic characteristics are in the nature of self-similarity. This investigation also describes a narrative method called a percentile method since it is percentiles for measuring Hurst exponent; it is a sign of the effectiveness of self-similarity.

(Sarla et al., 2016) focuses on studying and estimating self-similarity using only some Hurst parameter methods, which shows the intensity of self-similarity. Real-time data from various web centres deliberate here as traffic and examined by different procedures for testing self-similarity. The outcome of the experiment showed and verified that the traffic tested in this learn to be Self-similar by an innovative method percentile, supported on a few expressive methods.

(Chen Pu, 2017) generated time series under the FARIMA model, then using the R/S analysis method, measured Hurst index for the series. They showed that the generated series using the Hurst index is different from the Hurst index calculated for real sequence by the R/S method.

(Yu and Petropulu, 2004) studied the broadcast of self-similarity as wire line traffic is sent through a gateway to a wireless network. They utilize a generally used model for repacking and buffering at the gateway and compute the output traffic statistics. Study and simulations expose that the buffer system can build traffic with various degrees of self-similarity as balanced to the arriving traffic, yet traffic to no longer self-similar.

(Pushpalatha et al., 2016) reported that real-time internet users input data traffic exhibits self-similarity, also measured by different techniques to test self-similarity. Experiments approved that a web user's influx pattern has a Self-similar structure. Also, the average waiting time distribution and queue length at an internet center have been calculated against traffic intensity.

(Mallikarjuna Reddy and Pushpalatha Sarla, 2017) They presented a linear regression model fit in the direction of self-similarity behaviour of internet user's influx data pattern. The present study provides a mathematical model equation for linear regression to predict Internet users' arrival information at web centres.

(Joarder, 2008), discussed and characterises network traffic such as compressed audio and video. The data source is also inclined to make congestion traffic models and demonstrate the sure extent of correlation between arrivals and LRD in time streams to examine network performance. In this study, the real-time video and audio flowing traffic, collected as of the video conferencing session of an enterprise high-speed hybrid multimedia network, has been completed in the framework of self-similarity.

(Alberto Montanari et al., 1997) considered a FARIMA model to represent the combined effect of Short and Long memory in Hydrolic data. They presented a framework for the recognition and development of the method. They applied the model to Italy's Lake Maggiore to estimate daily and monthly inflows.

(Christos Katris and Sophia Daskalaki, 2011) compared ARIMA, FARIMA and ANN models by fitting to different time series data from public domain discussed characteristics such as LRD and self-similar nature in internet congestion. They expected FARIMA to dominate ARIMA when time series has the characteristics of self-similarity and LRD. They also mentioned that by identifying the existing non-linear structure in time series, predictions could be improved using ANN. Finally, they measured the forecasting efficiencies of the models by MAE and RMSE.

(Omekara C O et al., 2016) suggested ARFIMA develops an approach for forecasting commercial banks' liquidity ratio of Nigeria. They used the Augmented Dickey-Fuller test to check stationarity and applied the Hurst index to verify long-term memory characteristics. For the same data, they fitted a suitable ARIMA model. The best model was identified using AIC and RMSE and concluded that the ARFIMA approach was best in that regard.

(Manohar et al., 2019) studied the Self-similar behaviour of air passengers arrival pattern of Air India domestic services. They discussed various methods to calculate the Hurst index. The intensity of self-similarity also presented the importance of self-similarity from various aspects and concluded that Air India domestic air traffic has a Self-similar characteristic.

(Sumalatha et al., 2019) developed a forecasting architecture to forecast short term road congestion using ANNs; data considered was from a busy junction, Hyderabad, India, for peak hours in the morning for five days. MLP network employed to identify the network discussed the significance of the study in the Intelligent Transport System (ITS).

1.3 AIM AND SCOPE OF THE INVESTIGATION

In practical, most of the traffic is Self-Similar or Long Range Dependent (LRD), such as internet congestion, LAN, WAN etc., it is important to forecast the traffic accurately. However, classical time series forecasting models describe only short range dependence, and it is challenging to forecast the Self-Similar and LRD time series because of its strong correlation. In this thesis, we investigated the Self-Similar characteristics of air traffic, analyzed few existing forecasting models and compared their accuracy.

The principal focus of the thesis is

i. To fit some existing time series procedures to air traffic.

ii. To illustrate the theory and characteristics of Self-Similar behavior of air traffic.

iii. To fit a forecasting model to the Self-Similar air traffic and compare this model with the existing forecasting models through numerical results.

Basic concepts are presented to give a clear view. Numerical results will be presented, and various performance evaluation measures will be used to compare various forecasting methods time series. The subject of this thesis is the number of passengers travelled by Air India and Indigo scheduled Domestic flights.

The forecasting models studied in the thesis may not be limited to air traffic but also to any traffic or time series as long as the nature of the time series is Self-Similar or LRD. The numerical results provide some insight into forecasting Self-Similar air traffic.

1.4 FRAMEWORK OF THE STUDY

Various forecasting models to air traffic data corresponding to Air India and Indigo Scheduled Domestic services have been studied. The forecasted models discussed were stochastic models, exponential smoothing models, ANN models. The self similarity of air traffic has investigated, and then the ARFIMA or FARIMA model was employed. This thesis contains six chapters, in addition to the list of references cited throughout the thesis.

Chapter I give the fundamentals of the domain of research. We discussed the role of Indian civil aviation along with its background and expansion. Industrial profiles of Air India and Indigo have been placed. A literature survey is made in this chapter.

Chapter II discussed basic concepts of time series analysis, preliminaries, and the concept of self-similarity introduced from various aspects.

In chapter III, various stochastic models and exponential smoothing models for forecasting the time series have been discussed. Appropriate stochastic model and Holt-Winters additive model have been fitted to Air India, and Indigo scheduled domestic air traffic. The best model has been identified using performance evaluation metrics. Numerical results, the plots exhibiting the relation of actual, predicted and forecasted are presented in the chapter.

In chapter IV, recent popular approach Artificial Neural Network (ANN) architecture to forecast the time series has been discussed. Various ANN models using Multi-Layer Perceptron (MLP) have been developed for Air India and Indigo Scheduled Domestic air traffic. The plots of actual and predicted, actual versus predicted, actual and forecasted and residuals by the developed model are presented in the chapter.

In Chapter V, we discussed measures for finding Hurst exponent of self-similar progression. Here we proposed a model depend on Percentiles and conventional by existing methods such as Rescale Adjusted Range Statistics is projected. In this method, the Hurst index is approximated by measuring the mean of rescaled range data. Also, the Variance-Time method and that of its aggregating a process for self-similarity are briefed. Estimating the spectral density that is a series with LRD showed as Peridogram method. In time series analysis, classification of the predictable correlation be able to specified in terms of ACF is generalized Correlogram method is used to fit the self-similar process. Higuchi's method proposed is related to the absolute moments used to show self-similarity and apply the residual of the regression method. Study on Self similar behaviour of Air India and Indigo air traffic is discussed using Hurst index by three methods Correlogram, Variance Time method and Percentile method. ARFIMA (FARIMA) model has been fitted to Air India and Indigo air traffic. The plot of actual and forecasted values by the model is presented. Overall comparisons have been presented at the end of the chapter. ARFIMA is compared with ARIMA, SARIMA, ANN and Holt-Winters additive models.

In Chapter VI, summery, overall conclusions and possible open problems are placed. At the end, a list of references mentioned at various places in the entire thesis has been placed.

1.5 INDUSTRIAL PROFILE

Aviation in India broadly classified into two categories, Civil as well as Military. IATA (International Air Transport Association) reported that Indian aviation is speedily heightening aviation trade globally. Indian civil aeronautics market is the world's 3rd massive civil aeronautics market. The growth of civil aviation connectivity and infrastructure in India is driven by the Ude Desh ka Aam Naagrik (UDAN) scheme. UDAN-RCS is an Indian government scheme. UDAN is progress of airport regionally as well as RCS (Regional Connectivity Scheme). This project intends to allow the nation's ordinary citizen to fly that aims to arrange air journeys that are inexpensive and widespread to encourage overall countrywide.

1.5.1 Aviation Economy

Indian Civil Aviation market is the 3rd massive civil aeronautics trade in the globe. In 2016, the Indian civil aviation market displayed 131 million travellers of air traffic, among them 100 million passengers travelled by domestic services. In 2020 the commerce was expected to own 800 aircraft.

1.5.2 Organizations

- **MOCA (Ministry of Civil Aviation)**
- **DGCA (Directorate General of Civil Aviation)**
- **AAI (Airports Authority of India)**

1.5.3 Airports

India has three advanced airports in terms of parallel runways; they are Hyderabad, Delhi also Bangalore. There are 132 airports as of November 2014 in India. Among them, most of the air traffic handled by the airports in Bangalore, Delhi, Kolkata, Mumbai, Kochi, Jaipur, Pune, Chennai, Ahmadabad and Hyderabad.

1.6 SUBJECT OF THE THESIS

In the present thesis number of passengers is treated as air traffic. We consider the number of passengers travelled monthly by **Air India** and **Indigo** scheduled Domestic services from 2012 January to 2018 December. The details were captured from the official website of DGCA (Directorate General of Civil Aviation).

1.6.1 Air India

Air India is the Indian air transporter service. In 1932, the airline was initiated by Tata in the name Tata airlines. He travelled through the air a consignment of airmail by its basic one engine de Havill and puss motho from Karachi to Juhu aerodrome, Mumbai, then continued to Madras. Later than Second World War, Tata airlines have changed the name to Air India also transformed into a public limited corporation. It is the first Asian airline that introduced a jet airliner in its fleet. On 1960, February 21, Air India grabbed dispatch of its primary jet aircraft Boeing 707, labeled as Gauri Shankar.

New Delhi is the headquarter of Air India. It is a Government enterprise owned by Air India Limited. It was formed by merging Indian airlines as well as Air India in 2007. The Air India Limited set off as the 27th Star Alliance member in 2014. IT is serving 94 destinations, including 40 International and 54 Domestic stations in 28 nations covering four continents on every side of the globe. It has its main pivot at New Delhi, Indira Gandhi international airport, secondary hub at Mumbai, Chatrapathi Shivaji International airport, various focus cities in India. It is India's largest International airliner having 18.6% of the market portion. Air India got approval for privatization by the Government of India on 28th June 2017. A committee was set up, and the process was started.

1.6.2 Indigo (Inter Globe Aviation Limited)

Indigo is economical air transporter of India, and its headquarter is at Gurgaon, Haryana, India. In India, Indigo is the vast air transporter with regard to fleet size and travelers transported. As of August 2019, it has a Domestic market share of 47.0%. Indigo is the seventh massive air transporter in Asia carried 46 million above travelers in 2017. Also, Indigo is the massive independent economical air transporter in Asia by passengers carried and jet fleet size.

The Indigo originated by Rahul Bhatia of Inter Globe Enterprises and Rakesh Gangwal, a US-related expatriate Indian, as a private corporation in2006. In July 2006, it got delivered its first aircraft and started operations after a month. In 2012 it became the immense Indian transporter in the traveler market portion. The airline is serving 83 destinations, including 23 International and 60 Domestic stations. Its main hub is at Delhi, Indira Gandhi International airport.

CHAPTER II

CONCEPTS OF TIME SERIES ANALYSIS AND SELF-SIMILARITY

2.1 DEFINITION OF A TIME SERIES

Time series is a sequence of observations collected in chronological arrangement. It is denoted by $\{Z_t\}$, $t_1, t_2, \dots, t_n$ be n time points which are equidistant and $Z_1, Z_2, \dots, Z_n$ be the corresponding values of some study variable. Z(t) is treated as a random variable, t can be years, months, weeks, days, hours and even seconds.

A time series collected for one variable is termed as univariate time series, whereas multivariate time series is for multiple variables. A time series may be discrete or continuous. When a time series considers values at discrete time points is a discrete time series, and when it considers values at a continuous time interval, the time series is continuous. The number of vehicles passing in a particular road hourly, numbers of air passengers travelled by an air carrier monthly, and India's population in every census are examples of discrete time series. Some of the continuous time series is the water level of a dam observed every hour, daily temperature of a particular region, yearly production in paddy.

2.2 COMPONENTS OF TIME SERIES

The different influences that cause the fluctuations in a time series are broadly classified into four groups, usually known as time series components. They are Trend, Seasonal, Cyclic and Irregular or Random variations. The four factors are briefly described below. When observed for a long period of time, the characteristic revealed by the time series to rise or decline is termed as a trend. For example, population growth, literacy rate, and monthly expenditure of a family show an upward trend, whereas mortality rates and labor force engaged in agriculture show a downward trend.

Seasonal variations are periodic in nature and repeatedly occur in regular intervals of time within a period of one year or less. The main factor of seasonal variations is due to Natural (climate and weather conditions) and manmade conventions (festivals, conventions, customary practices etc.). Deals of air coolers, ice cream raise in a warm season, demand for gold ornaments raise in festivals, marriage seasons.

Cyclical variations are also periodic in nature and consist of oscillatory movements (up and down). Cyclical variations extend above a long spell of time, generally several years; they cannot be predicted to occur with a regular period of times. Cyclical variations may be present in the economical and financial time series data. In general, a business cycle has four phases- i) Prosperity or period of boom ii) Recession iii) Stagnation or Depression iv)Upturn or Recovery.

The Random or Irregular fluctuations are not caused by the trend, seasonal and cyclical variations are called Irregular fluctuations. They are unpredictable and beyond the control of the human hand. Some irregular causes of variations are wars, floods, earthquakes etc.

2.3 DECOMPOSITION OF TIME SERIES

To study the effect of four factors generally, two decomposition methods are applied in time series.

Additive model

$$Z_t = T_t + S_t + C_t + I_t \tag{2.1}$$

Where T_t is the effect of trend, S_t is the effect of seasonal variations, C_t is the effect of cyclic variations, I_t is the effect of random variations observed at time t, in this model, the components are independent of each other.

Multiplicative model

$$Z_t = T_t * S_t * C_t * I_t \tag{2.2}$$

This approach assumes that the elements need not to be independent.

Time series data generally be viewed in almost all the fields such as economic business, engineers, geography, meteorology and the social sciences. Few examples of time series plots are shown below.

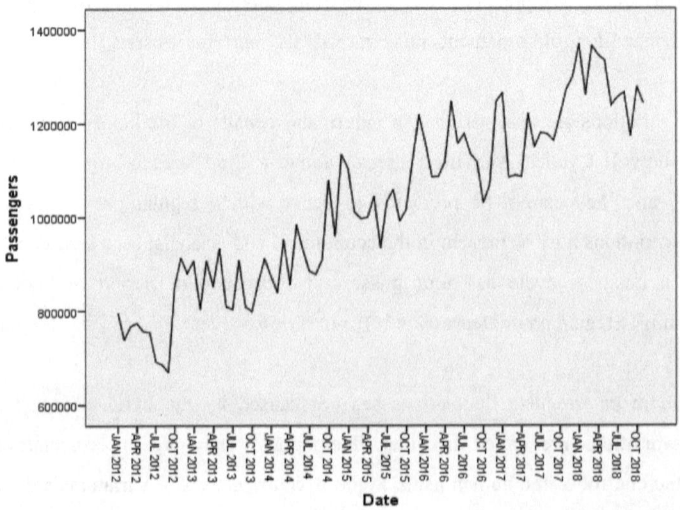

Figure2.1 Number of passengers travelled monthly by Air India Domestic flights.

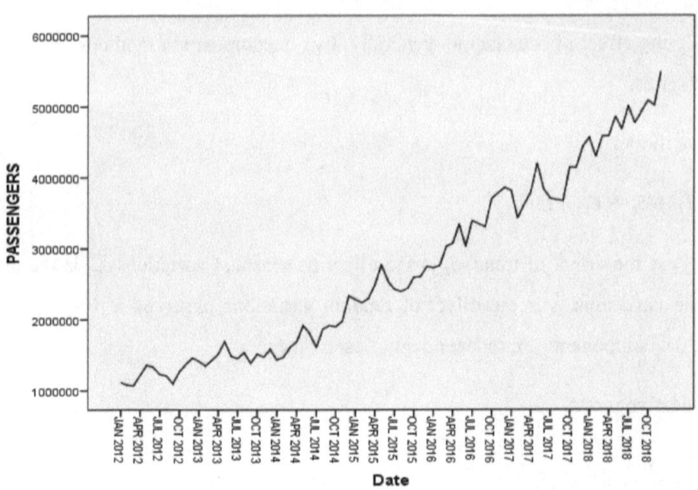

Figure 2.2 Number of passengers travelled monthly by Indigo Domestic flights.

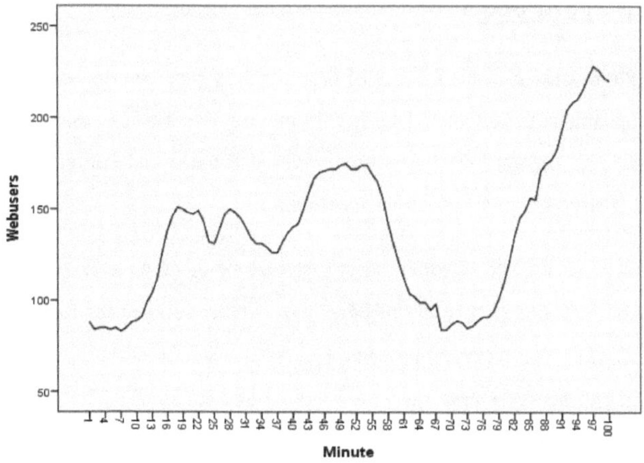

Figure2.3 Count of Web users signed into an Internet Server per Minute

Figure 2.1 and 2.2 are the plots of the number of passengers travelled monthly by Air India and Indigo Domestic flights, respectively and figure 2.3 represents the count of Web users signed into an Internet Server per Minute.

2.4 ANALYSIS OF TIME SERIES

An essential feature of time series is that successive observations are dependent. This dependence among the observations is practically important in any practical situation. The exploration of the dependence with time series techniques is time series analysis. This involves the development of stochastic and dynamic structure and applications of those models in particular areas (Hipel et al., 1994). Forecasting future values using previous and present values using developed models of time series is the main focus of time series analysis. Developing a proper time series model is very important, which results in an accurate forecast.

The probabilistic design of the sequence of data points described by a model is called a stochastic process (John H Cochrane, 1997). An N successive data points $Z_1, Z_2, \ldots, Z_n$ from a time series is a sample from the stochastic process. The main desire of statistical analysis is to conjecture characteristics of the universe from the characteristics of the sample. To forecast means to deduce the probabilistic pattern of future values from the universe under specified sample Z_t of previous data points.

2.5 STATIONARY PROCESS

A key aspect in the invention of the models of time series is based on the supposition that it is of the appearance of statistical stability. In precise, this kind of assumption is stationarity. In general, the stationarity of time series defined by its variance, mean and autocorrelation structure or spectral density function.

$\{Z_t, t \in T\}$ be a time series and is called strictly stationary of order n (n ≥ 1) if the joint probability density function (jpdf) of $\{Z_{t1}, Z_{t2}, \ldots, Z_{tn}\}$ is same as the jpdf of $\{Z_{t1+h}, Z_{t2+h}, \ldots, Z_{tn+h}\}$, for every values of $t_1, t_2, \ldots, t_n$, h & n. In other words, if the jpdf is time invariant, the time series $\{Z_t, t \in T\}$ is strictly stationary. This is a very strong condition which states that Z_t's are identically independently distributed (i.i.d) (John H Cochrane, 1997 and K W Hipel et al., 1994).

But in materialistic implementations, the supposition of strictly stationary is not necessarily needed; therefore, a weekly or second order or covariance stationarity is considered. A time series is stationary if variance and mean are time independent. Stationarity in any time series data can be tested by Dickey and Fuller John (H Cochrane, 1997) etc.

In practical applications, time series data may be non stationary. As per Hipel and McLeod, as the time series observations increase, the chance of non-stationarity of time series increases. When the data is not stationary, it is to be converted to stationary by stabilizing variance and mean of the series. Preferably variance to be stabilized first and then stabilize means. The variance of a time series can be stabilized by using square root or logarithmic transformations successively. The mean of data can be made stable by successive differencing.

It is to be noted that one observation will be lost in each differencing. Normally it requires one or two differencing to stabilize the mean. By stabilizing mean we eliminate the trend in the data.

Sometimes seasonal patterns may also be present in the practical data. Seasonal patterns may be removed by using seasonal differencing. If there are s seasons in a year, then $W_t = Z_t - Z_{t-s}$ generally there will be 12 seasons for monthly data and 4 seasons for quarterly data.

The stationarity condition also states that the covariance of Z_t, Z_{t+k} segregated by k time intervals should be the same for all t. The covariance between Z_t and Z_{t+k} for lag k is called the autocovariance of lag k and is given by

$$\gamma_k = cov(Z_t, Z_{t+k}) = E[(Z_t - \mu)(Z_{t+k} + \mu)] \quad (2.3)$$

A detailed discussion on ACF and ACVFs are presented in the next chapter.

2.6 PRINCIPLE OF PARSIMONY

The mathematical models developed for the time series data contain certain unknowns of the population to be computed from the sample. In developing a proper time series model, the principle of parsimony places a vital task in the selection of parameters. This principle says to consider as less parameters as possible for requisite description. Our objective is to develop an adequate but parsimonious model to forecast, attention and attempt is required in model identification. The selection exercise goes in an iterative way; initially, one should consider the simplest model out of a number of suitable models.

2.7 DEFINITIONS OF SELF-SIMILARITY

2.7.1 Overview

In this chapter, we started with the preliminaries of self-similarity and their pertinent issues. In the primary segment, we initiate the concept of self-similarity from various aspects; also present the models which generate self-similar traffic. A few of these techniques are based on Pareto distribution and Normal distribution, and some models involve time series. Finally, methods for estimating the Hurst parameter are discussed. The Hurst index value provides the intensity of self-similarity. Apart from the existing methods, we introduce a novel method called the percentile method to compute the Hurst parameter.

2.7.2 Self-similarity

The notion of self-similarity was pioneered (Mandelbrot, 1968; Willinger and Kihong Park, 2000) and is used to support modeling Geological and Hydrological problems.

Self-similarity is a word that is a convinced attribute of an entity that is retained w.r.t scaling in time or/and space. If an entity is self-similar, then its elements, as enhanced, bear a resemblance to the whole configuration.

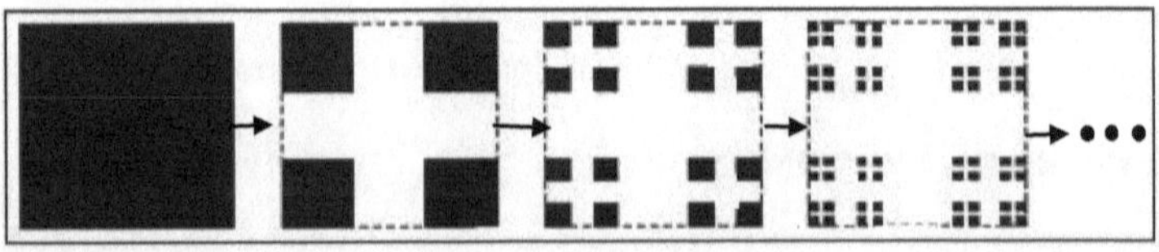

Figure 2.4: Contor set

For the situation of time series, Statistics are potted with respect to size and time. Another example is two dimensional Contor set which is put up as follows: Consider the set [0; 1] £ [0; 1] which is unit square. If each side is trisected, then nine smaller squares are formed. All the squares, except the square at the corners, are to be dropped. If the same process is repeated to get smaller squares, one obtains a limit set known as the Contor set. This productive process is depicted in Figure 2.4 Contor set has good applications in Science and Engineering.

2.7.3 Self-similarity screening in the natural world

In general, self-similarity occurs when the shape of a thing is similar or approximately the same as the part of itself. That is, every portion of the self-similar object can be considered a reduced scale of the whole. In natural science, there are a lot of examples of self-similarity, such as fern, snowflake, mountain ridges, coastlines etc. From a view of statistics, self-similarity is defined as two subsets of the whole set that are invariant in statistical distribution on different scales. In other words, self-similarity implies being rescaled or translated on the original signal keeps the same statistical distribution.

As a fresh and interesting topic, many studies on self-similarity have been done in different fields, such as Magneto Hydrodynamics (MHD) in physics, the human nervous system in biology, stellar fragments in astronomy and option pricing in finance etc.

In historical mathematics, the idea of self-similarity was first introduced (Mandelbrot, 1968). Self-similarity is so broad and plays a significant role in nature, society, and arts. It can only be no longer computed without as a minimum theoretical indulgent of this barely credible phenomenon. Fortunately, the main thoughts of self-similarity are not so complicated to skilled and illustrations of self-similarity are extraordinarily simple to find out. We can merely understand the fundamentals of self-similar phenomena and be able to recognize them practically everywhere we look.

When graphically mapped, internet traffic provides a striking example of self-similarity in the technological world. In this fractal model, internet hubs (Google and Face book) are represented as the largest nodes on the graphic. Smaller nodes branch off of the major ones, and these represent popular but slightly less travelled sites. And this pattern repeats. At the furthest reaches of the graphic are local networks that receive very little daily traffic. So, basically, based just on graphic representations, our brains, the universe, and the internet are all the same thing.

In the research ground of music information recovery, self-similarity generally refers to the fact that music repeatedly consists of parts that are frequently in time. In other words, music is self-similar under chronological conversion rather than under scaling. In music, strict canons show various amounts of self-similarity, as perform segments of fugues. Sheppard tone is self-similar in the frequency domains.

Romanesco broccoli shows sign of strong self-similarity. Broccoli is another kind of cauliflower. It exhibits strong self-similarity that is a tedious fact. Romanesco broccoli is, too possibly the most extremely fractal vegetable in the world. Its pattern absolutely represents with spirals upon spirals of the vegetable pattern replicating themselves at various levels of scale.

Figure 2.5: Dill plant

The Dill plant gives a pleasing and simple example of self-similarity with two scales of structure. In the Figure 2.5, the main stem goes up to a joint, as of which various secondary stems branch out. Towards the end of every second level stem, there is a small copy through shining third level stems, so on present the impressive example of self-similarity. Various categories of cactus plants, for example, the Saguaro, have enlargement patterns that turn out a ladder of self-similar lobes.

Having a fairly broad sampling of self-similarity in different areas of nature, make a decision to attempt something new by looking for more conceptual models of this design in common systems, arts and mathematics. If one supposes how our governments, legal systems, different types of economic systems, or law enforcement agencies effort, they have quite discrete hierarchical collections with central, state and neighbouring levels frequently being the most important scales. Fundamentally a similar kind of movement is happening at various scales, and thus we discover a well-known design within a new background.

The main characteristic of self-similarity is the same thing on different scales can be obtained in innumerable examples from the social divisions. (Chin Wen Cheong, 2010) proposed the self-similarity through fractionally integrated techniques in financial markets. Self-similarity has vital effects intended for computer networks like distinctive network traffic has shown the self-similar properties.

For instance, packet switched data patterns, and telegraphic engineering traffic flows seem to be present as statistically self-similar. This possession indicates that simple techniques using a Poisson distribution are imprecise networks intended without enchanting self-similarity into concern is probably to function in an unpredicted manner. In the same way, stock market associations are illustrated as exhibiting self-affinity; that is, they come into view as self-similar when transformed through a suitable affine transformation intended for the exposed level of aspect.

Finite subdivision regulations are powerful techniques used for structuring self-similar sets, including the Sierpinski triangle and the Cantor set Well, we might continue with additional examples of self-similarity, such as designs on shells, lightning bolts, metal ions or aggregation of bacteria, crystallization patterns in agate, surfaces of cancer cells, quantum particle paths, scores of scaling laws in biology, gamma-ray burst fluctuations, abundances or species distributions, renormalization in quantum electrodynamics, drop formation and so on. We will have various opportunities in the future for our compilation. Since, further, than any reasonable hesitation, nature is passionate about self-similarity.

2.7.4 EVIDENCE OF SELF-SIMILARITY IN NETWORK TRAFFIC

(Leland et al., 1994) made a seminal study at AT&T Bell Labs, USA and measured real Ethernet traffic. The pertinent observations are depicted in Figure 2.6 from this figure; the following inferences can be made. In the case of Ethernet traffic, no matter what time scale we choose, we shall see similar patterns, unlike the other one. In the case of Ethernet traffic, burstiness exhibits across all time scale, unlike the other one. Hence, network traffic modelling that uses the Poisson process does not take self-similar characteristics in data congestion into consideration. The observable fact of traffic self-similarity has a significant blow on queuing concert and is obtained a major concentration in the network exploring society.

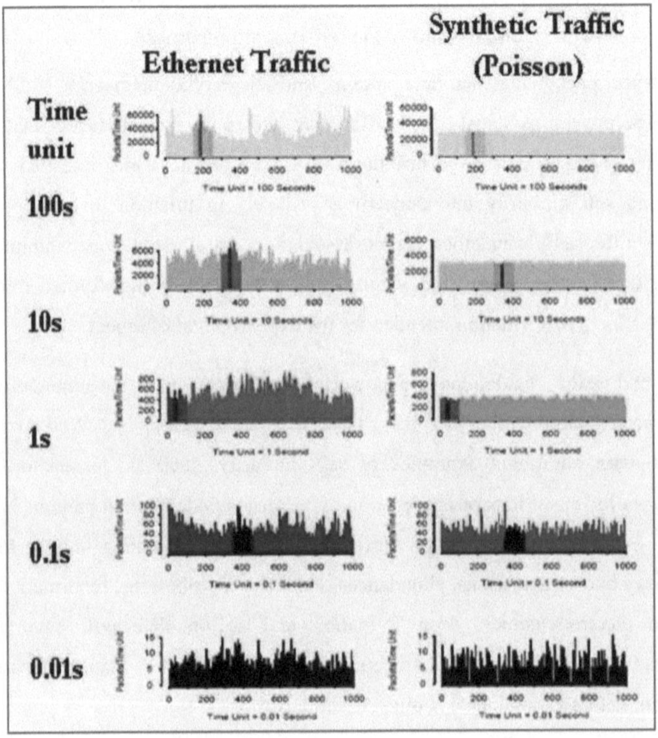

Figure 2.6: Pictographic proof of self-similar traffic (Ethernet traffic measured at AT&T Bell Labs, New Jersy, USA)

2.7.5 PROPERTIES OF SELF-SIMILARITY

(Mandelbrot, 1968) invented the word self-similar. They obtained self-similar progression for the awareness of statisticians, generally as functions in some regions like geophysics and hydrology.

When the procedure is analyzed at unreliable time scales, Self-similar processes remain unaffected by the Probabilistic properties. This is during dissimilarity for Poisson processes which mislay their squash out with burstiness as time scales be altered, despite the fact that a time series displays burstiness in excess of a broad range. Self-similar processes emerge in different parts of probability theory, such as statistical physics, fragmentation theory, random fields.

Self-similarity has been recognized as a typical property of objects introduced as counter exceptions to proposed general rules in mathematical analysis and considered now as early examples of fractals. An entity or object is compiled into subunits then sub-sub units on multiple stages that statistically are similar to the entire object's formation. Mathematically this property should embrace on all time scales. In real life situations, there are unavoidably lower and upper bounds over which h such self-similar nature apply. Scale invariance is a fundamental property of ensembles of natural images.

Self-similarity and fractals are intimately related (Peitgen, 1992) some of the defining properties of fractal processes are as follows:

i. Performance of Self similar
ii. Characteristic of Long-range dependence
iii. Variance decays Slowly
iv. Distributions with Heavy tails
v. 1/f noise
vi. Dimensions of Fractal
vii. SRD verses LRD

i. Self-Similarity

As mentioned above, a procedure is well thought-out on the way to exist precisely Self-similar, if collective methods have first-order and second-order statistical properties; so as to they are impossible to differentiate from those of the self-similar process itself. On the contrary, the ACF of the aggregated method move towards that of the method itself intended for a large extent of aggregation in an asymptotically self-similar process.

When a self-similar progression X_t has steady enhancements, the procedure may be a fundamental process deferring the fractal process. That means one can construct a stationary augmentation process.

$$X_n = X[nT_i] - X[(n-1)T_i] \qquad (2.4)$$

With LRD, 1/f - noise and slowly decaying variance properties, which are specific for fractal processes. The most extensively-recognized pattern of the self-similar process is an FBM process (with infinitely long-run correlations), a generality of Brownian motion with uncorrelated and independent intervals (Beran, 1994).

ii. Long-Range Dependence

In a self-similar process, LRD refers to the determination phenomenon, and it is primarily expressed in stipulations of the performance of the auto covariance function. If X is a process supposed to exist Long-Range Dependence, and for fixed *Ti*, the auto covariance function to be non- summable.

$$\sum_{k=0}^{\infty} Cov(k) = \infty \tag{2.5}$$

It is observed that this definition discloses little information about the process X as it depends only on sum of the auto covariance function, not on an exact form. An enhanced definition is the one that denotes the behaviour of the tail of the auto covariance function, namely Long-Range Dependence has an Auto Covariance Function so as to perish hyperbolically.

$$Cov(X_n, X_{n+k}) \sim |k|^{-\beta} \tag{2.6}$$

As $|k| \to\infty$ and $0 < \beta < 1$. The β parameter described by Hurst Parameter

$$\beta = 2(1-H) \tag{2.7}$$

In distinguish, the SRD process has exponentially decomposing auto covariance.

$$Cov(X_n, X_{n+k}) \sim a^{|k|} \tag{2.8}$$

When $|k| \to\infty$ also $0 < a < 1$. The result displays that Short-Range Dependence processes have summable auto covariance

$$\sum_{k=0}^{\infty} \text{cov}(k) = \textit{finite} < \infty \tag{2.9}$$

The conclusion is that, for all the large values of k, though individual values of Cov (k) are small, their collective result be considerable lest of Long-Range Dependence processes and generates attributes to significantly dissimilar from those of Short-Range Dependence processes.

iii. Slowly decaying variance

Since stated above, the variance of sample mean reduces slowly but surely than the inverse to the size of the sample of a self similar procedure.

$$Var[X^m] \sim m^{-\beta} \tag{2.10}$$

Adequately for large m. with SRD, the parameter

$$\beta = 1 \text{ with } \quad Var[X^{(m)}] \sim m^{-1} \tag{2.11}$$

Property that the variance decays slowly by gradually-decomposing may exist and simply noticed by describing $Var\left[X^{(m)}\right]$ versus m lying on a plot on a log-log scale. A trend equation having a slope less than one specifies that a gradually decomposing variance. This property is able to perceive by an assist of the indicator of variation for enumerations.

$$F\,(T) = \frac{Var\,[N(T)]}{E[N(T)]} \tag{2.12}$$

$$Var[X^m] = \frac{E[N(mT_i)F[mT_i]}{m^2} \tag{2.13}$$

Entailing that moment of variance in time plot characterizing the gradually decomposing variance and with parameter β is corresponding to an Index of Dispersion for Counts with 1-β is the slope. In statistical tests, an assumption of slowly decaying variance property denotes that the normal measures of standard deviation σ are incorrect through an aspect when the size of the sample increases that reaches to infinity.

iv. Heavy-tailed distributions

In difference to the stated properties, which concern of time dependent scaling properties (like auto covariance function), the heavy-tailed distribution properties focus on the marginal distributions of X_n as described in Equation (2.4), which shows examples of ON/OFF periods and X_n are packet arrivals. The category of methods described in this mode is so far additional comprehensive than above stated compilations.

When the Complement of Distribution Function decomposes according to power law then X, the stochastic variable is known to have a heavy tail that is of the form.

$$\Pr\{X \geq x\} \sim x^{-\alpha} \text{ as } x \rightarrow \infty, \alpha > 0 \quad (2.14)$$

The main examination is that the property of distribution with heavy tails is not an essential state of self-similarity. Alternatively, self-similar characteristics of frequent traffic data outline results explicitly as of heavy tailed distributions.

v. 1/f - noise

This property is a frequency-domain expression of Long-Range Dependence. Particularly, the PSD of

$$S(w) = \sum_{k=-\infty}^{\infty} R(k) e^{-j2kw} \quad (2.15)$$

Long-Range Dependence processes diverge at low frequencies. It is in difference to Short-Range Dependence processes where the spectral density remains flat at low frequencies. In other words, the PSD of LRD progressions explain the power-law behaviour close to the frequency origin

$$S(w) \sim \frac{1}{|w|^{\gamma}} \quad (2.16)$$

When $|w| \rightarrow 0$ and $0 < \gamma < 1$. γ is the index associated to β than

$$\gamma = 1 - \beta = 2H - 1 \quad (2.17)$$

when $|w| \rightarrow 0$, for SRD procedures, the PSD remains finite, that is
When $\gamma = 0$ indicates the values of autocorrelation function decay adequately used for large values of k on the way to forming a finite sum.

vi. Fractal dimensions

Let d is the fractal dimension is an entity be described as follows:

$$d = \frac{\ln N}{\ln \frac{1}{\eta}} \quad (2.18)$$

Where η is the linear scale of objects and N is the number of self-similar entities to enfold a specified d dimensional entity. In contrast toward Short-Range Dependence processes (that are structured through a fractal dimension d=1), self-similar processes able to model by processes with fractional dimensions (i.e., d has the form d = xx.yy …)

This is a direct consequence of the slow and very slow convergence of the process to mean values.

vii. SRD versus LRD

Mathematically, the variation among SRD and LRD process (Cox D.R., 1984) be able to affirmed as follows:
In the case of Short-Range Dependence processes, the following properties are characterized.

a. $E[X_n^m]$ approaches pure noise of second order as m →∞
b. For big m, Var[X^m] asymptotically will be in the structure $\frac{Var\,[X]}{m}$
c. $\sum_{k=0}^{\infty} Cov(X_n, X_{n+k})$is convergent
d. S(w) is a Spectrum be finite at w =0

Whereas Long-Range Dependence processes, the following properties are characterized.

e. $E[X_n^m]$, as m→ ∞ doesn't reach second degree pure noise
f. For large m, $Var(X^m)$ asymptotically is in the structure $m^{-\beta}$
g. $\sum_{k=0}^{\infty} Cov(X_n,\, X_{n+k})$ is divergent
h. S (w) is a spectrum singular at w =0

2.8 MATHEMATICAL DEFINITION OF SELF-SIMILARITY

Definition 1. A stochastic process {x (t), t≥ 0} is known as self-similar if whichever

a > 0, there subsists $b > 0$ such that

$$X(at) = d\{bX(t)\} \qquad (2.19)$$

Let d indicates the parity of limited dimensional distributions.

Definition 2. Arrival patterns are modelled as a point process. Assume that $X = \{X_t / \text{ t}$ = 1, 2…} are the arrivals in t^{th} interval, segregate the timeline into exclusive intervals of size one. Whether X be a second degree stationary process has ACF $\gamma(k)$; $k \geq 0$ and variance σ^2, presented as

$$\gamma(k) = \frac{Cov(X_t, X_{t+k})}{Var(X_t)} \tag{2.20}$$

The process 'X' is known to be perfectly second degree self-similar accompanied by the Hurst index H and variance σ^2if

$$\gamma(k) = \frac{\sigma^2}{2}[(k+1)^{2H} - 2k^{2H} + (k-1)^{2H}], \forall \mathrm{k} \geq 1 \tag{2.21}$$

Definition 3. Regard to every m (1, 2, 3,....), assume another time series as $X^{(m)}$={$X_t^{(m)}$/ t =1,2,...} is determined by taking the average of unique process X on exclusive blocks with length m. i.e.

$$X_t^{(m)} = \frac{1}{m}\sum_{i=1}^{m} \mathrm{x}_{(t-1)m+i}, \mathrm{t} = 1, 2, ... \tag{2.22}$$

For each m, this another set, $X_t^{(m)}$, is again a weak stationary procedure with ACF of $\gamma^{(m)}(k)$. Hurst index H and variance if the procedure 'X' is termed as asymptotically weak self-similar in terms of dispersion of the averaged process, we describe a similar process as

$$\gamma^{(m)}(k) = \frac{\sigma^2}{2}[(k+1)^{2H} - 2k^{2H} + (K-1)^{2H}], \forall k \geq 1 \tag{2.23}$$

Definition 4. The procedure 'X' is known as exactly second degree self-similar accompanied by Hurst index H equals to 1-$\frac{\beta}{2}$ and variance σ^2 if

$$\mathrm{Var}\ (X^{(m)}) = \sigma^2 m^{-\beta}, \forall \mathrm{m} \geq 1 \tag{2.24}$$

2.9 LONG RANGE DEPENDENCE (LRD)

The differentiation between SRD and LRD procedures, for $H \neq 0.5$, by the Equation (2.21), we get

$$\gamma(k) = H(2H\text{-}1)k^{2\mathrm{H}-2} \text{ as } k \to \infty, \tag{2.25}$$

and we have

$$\sum_k \gamma(k) \sim c \sum_k k^{-\beta}, c = H(2H-1) \tag{2.26}$$

The series $c\sum_k k^{-\beta}$ is not convergent if $0 < \beta < 1$ (or) $0.5 < H < 1$ or else they are convergent actually a positive series. Consequently, the LHS $\sum_k \gamma(k)$ is divergent if $0 < \beta < 1$ (or) $0.5 < H < 1$ or else they are convergent. The ACF decays gradually, that is hyperbolically for $0.5 < H < 1$. In such case X is known as LRD, otherwise it is SRD for $0 < H < 0.5$, in this case the ACF is summable.

2.10 FIRST AND SECOND ORDER STATIONARY PROCESSES

2.10.1 First order stationary process

Let X is a stationary process of degree one when the first order density model of it doesn't vary with an operator time Δ. that is

$$f_X\,(x_{1,}t_1) = f_X(x_{1,}t_1 + \Delta),\ \forall t_1 \tag{2.27}$$

Now,

$$E[X_t] = \bar{X} = \text{constant} \tag{2.28}$$

2.10.2 Second degree stationary process

A procedure is called Second degree stationary process when first and second degree density function satisfies

$$f_X\,(x_{1,}t_1) = f_X(x_{1,}t_1 + \Delta), \tag{2.29}$$

$$f_X\,(x_1, x_2 : t_1, t_2) = f_X\,(x_1, x_2 : t_1 + \Delta,\ t_2 + \Delta) \tag{2.30}$$

for all t_1, t_2 and Δ.

2.11 MODELS ON DISTRIBUTIONS

2.11.1 Fractional Brownian motion

It was initially pioneered by (Mandelbrot and Van Ness, 1968) as a simplification of Brownian motion (Bm) (Mandelbrot et al., 1968). The Bm process $B(t)$ is the random function through the Gaussian increments, and where the average assessment of augmentation progression is nil.

$$E[B(t+\tau) - B(t) = 0 \tag{2.31}$$

Also, the variance is proportional to the time discrepancy.

$$\text{Var}\ [B(t+\tau) - B(t)] = \sigma^2 |\tau| \tag{2.32}$$

Alternatively, the fBm procedure B_H (t) is defined as dB (t) moving average, where the previous augmentations of B (t) are weighted through kernel $(t-\tau)^{H-\frac{1}{2}}$

The Weyls fractional integral $B(t)$ is used to state fBm process:

$$B_H(\mathrm{t}) = \frac{1}{\Gamma\left(H+\frac{1}{2}\right)} \int_{-\infty}^{0} \left[(t-\tau)^{H-\frac{1}{2}} - (\tau)^{H-\frac{1}{2}}\right] dB(\tau) + \int_{0}^{t} (t-\tau)^{H-\frac{1}{2}} dB(\tau), \tag{2.33}$$

Here the function $\Gamma(x)$ is stands for the Gamma function. FBM is a zero mean and non discrete self-similar procedure of Gaussian (B_H (t) : $t \geq 0$) and is described for $t \geq 0$ through the index $0.5 \leq H < 1$. All the finite-dimensional marginal distributions are Gaussian. The process value at time t can be intended with:

$$B_H(t) = Xt^H \tag{2.34}$$

Where X is the random variable and follows Gaussian distribution with location parameter zero and scale parameter one, as of self-similarity,

$$B_H(\mathrm{a}t) = a^H B_H(t) \tag{2.35}$$

For $H = 0.5$, the ordinary Bm process is obtained. The foremost statistical tests of the fractional Brownian motion process are (Beran, 1994)

- Mean: $E[B_H(t)] = 0$
- Variance: $Var[B_H(t)] = Var\,[X\,t^{H}] = t^{2H}$
- Auto correlation: $R_{BH}(t,\tau) = E\,[B_H(t)B_H(\tau)] = \frac{1}{2}[t^{2H} + \tau^{2H} - |t-\tau|^{2H}]$
- Stationary increments: $Var[B_H(t)\text{-}B_H(\tau)] = |t-\tau|^2$

The FBM model is broadly used in analytical and simulation studies regarding performance evaluation of the structures are driven by self-similar traffic (Willinger et al., 1996).

2.11.2 Fractional Gaussian Noise

The FBM of the stationary increment process X_H is known as the fractional Gaussian noise. This process was initiated (Mandelbrot and Van Ness, 1968).

$$X_H = (X_H(t) = B_H(t+1) - B_H(t); t \geq 0) \tag{2.36}$$

The FGN is a method through Gaussian distribution with location parameter μ and scale parameter σ^2, where variances of the squared mean of enhancements are correlative to variations in time. Also, the autocorrelation function is

$$R(t)=\frac{1}{2}(|t+1|^{2H} - |t|^{2H} + |t+1|^{2H}). \tag{2.37}$$

As lim t tends to ∞

$$R(t) \sim H(2H-1)|t|^{2H-2} \tag{2.38}$$

when $0.5 < H < 1$, fGn be exactly self-similar

2.11.3 The ON-OFF model

The method is at first pioneered by Mandelbrot (Mandelbrot, 1963), is depends on renewal-rewards processes, this model was restructured (Willinger et al., 1994) in the framework of data traffic. The collective traffic is manufactured by multiplexing the independent ON/OFF resources, where every resource exchanges among these states on definite time intervals. The produced data is self-similar if numerous re-established benefits procedures are overlying, the returns are limited to zero and unity, and the in between re-established times have heavy tails.

On the other hand, the overlying of several ON-OFF origins demonstrating the unlimited variance characterization (Noah effect) gives Self-similar collective congestion considering FBM while the number of sources produces large (Joseph effect) with Hurst index.

$$H = \frac{3-\alpha}{2} \tag{2.39}$$

The ON-OFF model is very attractive also simple as, for example, it offers an instinct into the self-similarity nature of Ethernet congestion. A workstation attached to an Ethernet network can be, in this case, considered as a renewal-reward process if the workspace is communicating over the network where the process is in ON-state; otherwise, it is OFF-state. The Ethernet congestion is the consequence of the collection of renewal-rewards procedures.

2.11.4 Random midpoint displacement (RMD) method

The accurate generation of FBM mark out is practically infeasible (owing to a large quantity of essential CPU time), a more competent estimate algorithm is the RMD method (Lau et al., 1995). A self-similar process is capable of being manufactured with increasingly subdividing an interval in excess of which a sample path is produced by the midpoint as of the observations of closing observations.

At every partition, a null location parameter Gaussian dislocation is utilized to find out model route value at the centre of the subinterval. The fundamental idea of the RMD structure is to enlarge the formed progression repeatedly by attaching the latest observations at the centre from the closing observations. Picture 2.7 summarizes clearly the way of working the RMD algorithm. Figure 2.8 exemplifies the initial three stages of the algorithm, foremost to the creation of the series ($d_{3,1}$, $d_{3,2}$, $d_{3,3}$, $d_{3,4}$). The motive of separating an interval (0, 1) to be build the Gaussian enhancements of X. By attaching offsets to central values creates the marginal distribution of absolute consequence Gaussian. Further comprehensive conversations of the RMD approach (Peitgen et al., 1992; Lau et al., 1995)

Step.1 If for time instance t, the process $X(t)$ is calculated in the middle of 0 to 1, subsequently set off by taking $X(0) = 0$, choosing $X(1)$ like an artificial random value as of Normal distribution with the location parameter is 0 and scale parameter

$Var\ [X(1)] = \sigma_0^2$.

followed by $Var\ [X(1)] - X(0)] = \sigma_0^2$ (2.40)

Step 2. Subsequently, $X\left(\frac{1}{2}\right)$ is made from the mean of $X(0)$ and $X(1)$, hence

$X\left(\frac{1}{2}\right) = \frac{1}{2}(X(0) + X(1)) + d_1$. The recompense d_1 is a GRN, it is supposed to multiply with a scaling factor $\frac{1}{2}$, with the mean is 0 and dispersion S_1^2 of d_1. Evaluate the vision of this stage, then follows the subsequent step shows in Picture 2.8.

For $Var[X(t_2) - X(t_1)] = |t_2 - t_1|^{2H}\sigma_0^2$ is correct, for $0 \leq t_1 \leq t_2 \leq 1$, it should be needed when $Var[X\left(\frac{1}{2}\right) - X(0)] = \frac{1}{4}\ Var[X(1) - X(0)] + S_1^2 = \left(\frac{1}{2}\right)^{2H} \sigma_0^2$.

Thus $S_1^2 = \left(\frac{1}{2^1}^{2H}\right)(1-2^{2H-2})\sigma_0^2$ (2.41)

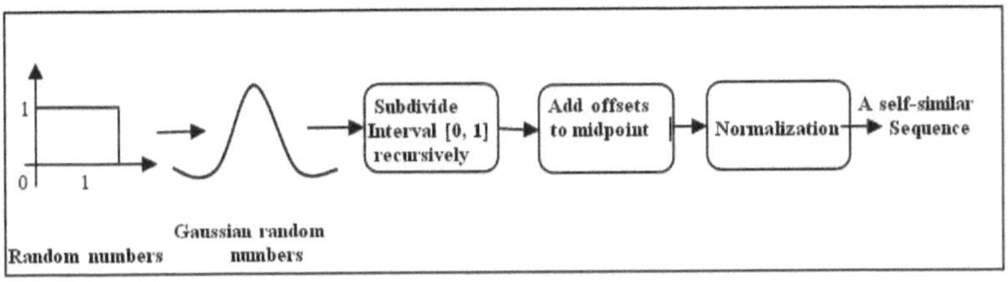

Figure 2.7: RMD approach

Step 3. Decrease the scaling element by$\sqrt{2}$ hence it is suppose$\frac{1}{\sqrt{8}}$, and split the intervals from 0 to $\frac{1}{2}$ and then $\frac{1}{2}$ to 1. The mean x($\frac{1}{4}$) is given as $\frac{1}{2}\left(X(0) + X\left(\frac{1}{2}\right)\right)$ plus an offset $d_{2,1}$,and is a GRN multiplied with current scaling factor $\frac{1}{\sqrt{8}}$. Then the resultant formula hold for $X\left(\frac{3}{4}\right)$,that is, $X\left(\frac{3}{4}\right)=\frac{1}{2}\left(X\left(\frac{1}{2}\right) + X(1)\right)+d_{2,2}$ is an offset worked out as before. So the variance S_2^2 of $d_{2,*}$; should be preferred such that

$$\mathrm{Var}\left[X\left(\frac{1}{4}\right) - X(0)\right]=\frac{1}{4}\mathrm{Var}\left[X\left(\frac{1}{2}\right) - X(0)\right]+S_2^2=\left(\frac{1}{2^2}\right)^{2H}\sigma_0^2 \tag{2.42}$$

Thus

$$S_2^2=\left(\frac{1}{2^2}\right)^{2H}(1-2^{2H-2})\,\sigma_0^2 \tag{2.43}$$

Step.4 In this step, continuing the same approach, decrease the scaling

factor by $\sqrt{2}$ i.e., carry out $\frac{1}{\sqrt{16}}$. Then set

$$X\left(\frac{1}{8}\right)=\frac{1}{2}\left(X(O) + X\left(\frac{1}{4}\right)\right)+d_{3,1};\ X\left(\frac{3}{8}\right)=\frac{1}{2}\left(X\left(\frac{1}{4}\right) + X\left(\frac{1}{2}\right)\right) + d_{3,2} \tag{2.44}$$

$$X\left(\frac{5}{8}\right)=\frac{1}{2}\left(X\left(\frac{1}{2}\right) + X\left(\frac{3}{4}\right)\right)+d_{3,3};\ X\left(\frac{7}{8}\right)=\frac{1}{2}\left(X\left(\frac{3}{4}\right) + X(1)\right) + d_{3,4} \tag{2.45}$$

In every formula, $d_{3,3}$ is calculated like a dissimilar GRN multiplied with the present scaling factor $\frac{1}{\sqrt{16}}$. The subsequent steps work out x(t) at $t=\frac{1}{\sqrt{16}}, \frac{3}{\sqrt{16}}; \ldots \frac{15}{\sqrt{16}}$

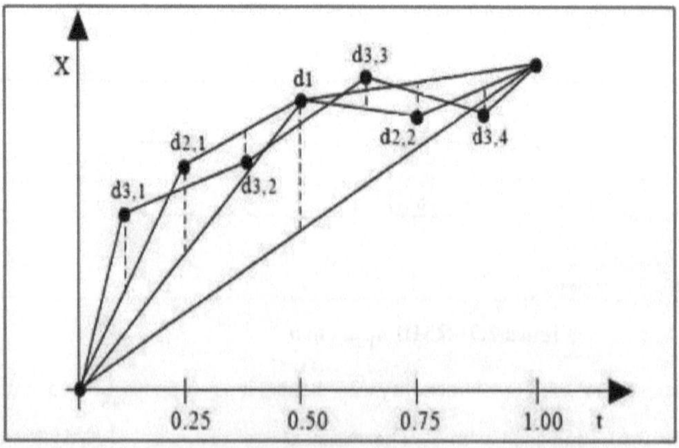

Figure 2.8: The first three steps of RMD method

By means of a scaling factor, another time decreased by$\sqrt{2}$ and persisted as specified above.

So the variance S_3^2 of $d_{3,*}$ must be selected such that

$$\mathrm{Var}\left(X\left(\tfrac{1}{8}\right)-X(0)\right)=\tfrac{1}{4}\,\mathrm{Var}\left(X\left(\tfrac{1}{4}\right)-X(0)\right)+S_3{}^2=\left(\tfrac{1}{2^3}\right)^{2H}\sigma_0{}^2. \qquad (2.46)$$

that is

$$S_2{}^3=\left(\tfrac{1}{2^3}\right)^{2H}(1-2^{2H-2})\sigma_0{}^2. \qquad (2.47)$$

The variance S_n^2 of $d_{n,*}$ therefore, yields$\left(\dfrac{1}{2^n}\right)^{2H}(1-2^{2H-2})\sigma_0^2$ (2.48)

The RMD method of the above 4 steps generates an estimated self-similar sequence for standard time series.

Chapter-III

ANALYSIS OF AIR TRAFFIC USING TIME SERIES MODELS

3.1 INTRODUCTION

In the second chapter, the fundamental notion of time series, the importance of time series analysis and forecasting has been discussed. The identification of a genuine time series design that exhibits the pattern of the time series under inspection is highly important. This selected design may be implemented for forecasting the upcoming observations. When the fitted model is adequate, the resulting forecast would be more accurate. Depending on the correspondence of the ongoing data point and the former data points, a time series model can be classified into linear and non linear models. Suppose the correspondence of the ongoing data point and the former data points exhibits as a linear function. In that case, the model is said to be a linear time series model, correspondence of ongoing data point and the former data points exhibits as a non linear function the model is said to be non linear time series model.

In time series analysis, the sequence may be represented by diverse forms of models that represent the different stochastic process. In a practical scenario, it is essential to develop frames that parsimoniously employ the parameters. Parsimony can be gainfully attained by representing the process with linearity with the help of less count of Autoregressive (AR) and Moving Average (MA) expression. Autoregressive (AR) and Moving Average (MA) frames (Box and G Jenkins, 1970, KW Hipel et al., 1994) are the immensely used linear models in time series analysis. A Union of Moving Average and Autoregressive model is the Autoregressive Moving Average (ARMA) (Box and Jenkins, 1970, John H Cochrane, 1997) model.

For non stationary data, Autoregressive Integrated Moving Average (ARIMA) is employed. The development to ARMA and ARIMA is Autoregressive Fractionally Integrated Moving Average (ARFIMA or FARIMA) (Park, 1999, Galbraith et al., 2001), allowing the differencing ‘d’ to take non-integer values.

A time series that has seasonal patterns can be analyzed by Seasonal ARIMA (SARIMA). The resource of all these designs is the principle of Box Jenkins; hence these models are called as Box-Jenkins models.

Linear models are extensively employed in the analysis of time series. They are accessible to apply also understand; apart from practical applications, some of the time series data may exhibit non linear relationship among the observations. So non linear time series models are used to analyze such data. Some of the non linear techniques are the Autoregressive Conditional Heteroskedasticity (ARCH) approach, GARCH (Generalized Autoregressive Conditional Heteroskedasticity), EGARCH (Exponential Generalized Autoregressive Conditional Heteroskedasticity) (H Park, 1999, R Parrelli, 2001) model etc.

A time series consists of a series of data points collected in accordance with the occurrence of time. The material collected over time consists of a few patterns with inherent non regular fluctuations. "Smoothing" is a widely used technique for reducing the effect of non regular changes. When this course of action is rightly employed, it clearly discloses the foundational Trend, Seasonal and Cyclic elements of the data.

Exponential smoothing is the most well liked proposal used to obtain a smoothed data set. This technique assigns exponentially weights for past observations in forecasting. Holt-Winter's Exponential smoothing proposals are popular in dealing with seasonal time series.

The present chapter discusses the linear time series procedures, exponential smoothing models and their important properties.

3.2 FORECASTING THE TIME SERIES BY STOCHASTIC MODELS

The first step in forecasting procedure is to check the stationarity of a time series. There are several tests for testing the stationarity. KPSS test is one such test procedure which is described below.

3.2.a. KPSS TEST FOR STATIONARITY:

The Kwiatkowski–Phillips–Schmidt–Shin (KPSS) test is used for testing the stationarity in a Time series with the Null Hypothesis: The time series is Stationary against the Alternative: The Time series is non-stationary. Contrary to most unit root tests, the presence of a unit root is not the null hypothesis but the alternative. Additionally, in the KPSS test, the absence of a unit root is not a proof of stationarity but, by design, of trend-stationarity. This is an important distinction since it is possible for a time series to be non-stationary, have no unit root yet be trend-stationary. In both unit root and trend-stationary processes, the mean can be growing or decreasing over time; however, in the presence of a shock, trend-stationary processes are mean-reverting (i.e. transitory, the time series will converge again towards the growing mean, which was not affected by the shock) while unit-root processes have a permanent impact on the mean (i.e. no convergence over time).

The KPSS test is based on linear regression.

$$x_t = r_t + \beta t + \varepsilon_1$$

It breaks up a series into three parts: a deterministic trend (βt), a random walk (r_t), and a stationary error (ε_t). The test uses Ordinary Least Squares to find the equation.

3.2.1 Partial Autocorrelation and Autocorrelation Functions (PACF and ACF)

Recognizing a suitable model for the data set, it makes important to estimate the Partial Autocorrelation Function (PACF) and Autocorrelation function (ACF) values. So linear stochastic models are developed based on PACF and ACF. These estimated values exhibit the relationship between the observations of a data set. Plots of PACF and ACF versus time lags are put in to regulate the degree of AR and MA models. These identified approaches are exercised to forecast the upcoming time series values. The definition of ACF and PACF are discussed below:

The ACF of lag k of a stationary time series $\{Z_t, t \in T\}$, is defined by

$$\rho_k = \frac{\gamma_k}{\gamma_0} \tag{3.1}$$

Where γ_k is the Auto covariance function of lag k given by

$$\gamma_k = cov(Z_t, Z_{t+k}) = E[(Z_t - \mu)(Z_{t+k} + \mu)] \tag{3.2}$$

And

$$\gamma_0 = V(Z_t) = V(Z_{t+k}) = E[Z_t - \mu]^2 \tag{3.3}$$

where $\mu = E(Z_t) = E(Z_{t+k})$.

The ACF of ρ_k is an even function that is $\rho_k = \rho_{-k}$, and also the limits of ACF (ρ_k) are -1 to 1, i.e. $-1 \leq \rho_k \leq 1$. It is possible to have two or more time series having the same ACF.

PACF of lag k is the correlation between Z_t and Z_{t+k} after removing the variable between Z_t and Z_{t+k} from the model, for lag 1, ACF is the same as PACF.

Generally, the stochastic process which produces the time series is not specified, obtaining the original ACF and PACF is unendurable. So they may be enumerated by the time series data under consideration. The estimate of Auto Covariance Function (ACVF) of lag k

$$\hat{\gamma}_k = c_k = \frac{1}{n-k}\sum_{i=1}^{n-k}(Z_t - \bar{Z})(Z_{t+k} - \bar{Z}) \tag{3.4}$$

Estimate of variance (γ_0) is given by

$$\hat{\gamma}_0 = c_0 = \frac{1}{n}\sum_{i=1}^{n}(Z_t - \bar{Z})^2 \tag{3.5}$$

Then the estimate of ACF of lag k is

$$r_k = \frac{c_k}{c_0} \tag{3.6}$$

The plot of sample ACF is used in identifying the degree of MA process and the plot of estimated PACF is beneficial in placing the degree of AR procedure. The sample PACF and ACF plots are presented at the end of the chapter for experimental datasets.

3.2.2 Random Process or White Noise Process

White noise procedure is an arrangement of independent random variables having mean zero and constant variance that is $E(a_t) = 0$ and $V(a_t) = \sigma^2{}_a$ for all t.

It is the simplest stochastic process. This a_t's used as errors in times series.

3.2.3 Autoregressive (AR) Process

In the Autoregressive (AR) Process, the current deviations from the mean μ is regressed on past deviations of the process. Let $\tilde{Z}_t = Z_{t-\mu}$ then the AR model is given as

$$\tilde{Z}_t = \emptyset_1 \tilde{Z}_{t-1} + \emptyset_2 \tilde{Z}_{t-2} + \cdots + \emptyset_p \tilde{Z}_{t-p} + a_t \quad (3.7)$$

$$\tilde{Z}_t = \emptyset_1 B \tilde{Z}_t + \emptyset_2 B^2 \tilde{Z}_t + \cdots + \emptyset_p B^p \tilde{Z}_t + a_t \quad (3.8)$$

$$\emptyset(B) \tilde{Z}_t = a_t \quad (3.9)$$

Where

$$\emptyset(B) = 1 - \emptyset_1 B - \emptyset_2 B^2 - \cdots - \emptyset_p B^p$$

is multinomial in B of degree p and is termed as an operator of Autoregressive (AR). $\emptyset_1, \emptyset_2, \ldots \emptyset_p$ are known as AR coefficients and a_t is a white noise procedure 3.2.2. B is the backward shift operator in the above equations.

It follows the following properties.

- $BZ_t = Z_{t-1}$
- $B^j Z_t = Z_{t-j}$

3.2.4 MA (Moving Average Procedure)

In the Moving Average (MA) Procedure, the current deviation of the process from mean μ is represented as a weighted aggregate of existing and previous data points from the random procedure a_t (3.2.2).

The Moving Average model is given as

$$\tilde{Z}_t = a_t - \theta_1 \, a_{t-1} - \theta_2 \, a_{t-2} - \cdots - \theta_q \, a_{t-q} \quad (3.10)$$

$$\tilde{Z}_t = a_t - \theta_1 B a_t - \theta_2 B^2 a_t \ldots - \theta_q B^q a_t \quad (3.11)$$

$$\tilde{Z}_t = \theta(B) a_t \quad (3.12)$$

Where

$$\theta(B) = 1 - \theta_1 B - \theta_2 B^2 - \cdots - \theta_q B^q$$

is multinomial in B of degree q and is termed as the operator of Moving Average (MA). $\theta_1, \theta_2 \ldots, \theta_q$ are known as MA coefficients.

3.2.5 Autoregressive Moving Average (ARMA) Process

For establishing parsimony, it is essential to mix Moving Average as well as Autoregressive components. Consequently, the spliced Autoregressive Moving average scheme can be utilized, which subsequently called as ARMA (p, q) scheme.

The ARMA model is specified as

$$\tilde{Z}_t = \emptyset_1\tilde{Z}_{t-1} + \emptyset_2\tilde{Z}_{t-2} \pm \cdots + \emptyset_p\tilde{Z}_{t-p} + a_t - \theta_1\, a_{t-1} - \theta_2\, a_{t-2} - \cdots - \theta_q \qquad (3.13)$$

or

$$\emptyset(B)\,\tilde{Z}_t = \theta(B)a_t \qquad (3.14)$$

Where $\emptyset(B)$ and $\theta(B)$ are polynomial in B of degree p and q, respectively.

3.2.6 Stationarity and Inevertibility Properties

AR (p) procedure is stationary if the roots of $\emptyset(B) = 0$ fall outer the unit circle, and this procedure is invertible every time. MA (q) scheme is every time stationary, and it is invertible when the roots of $\theta(B) = 0$ fall outer the unit circle. The scheme of ARMA (p, q) is stationary if the roots of $\emptyset(B) = 0$ fall outer the unit circle, and it is invertible if the roots of $\theta(B) = 0$ fall outer the unit circle.

3.2.7 Procedure of Autoregressive Integrated Moving Average (ARIMA)

The scheme of ARMA is used to analyze time series data which is stationary, but most of the empirical time series may not be stationary since they have no fixed mean, for example, stock price data, socio economic series etc. Such non stationary data would be analyzed by the ARIMA process, in which first it translates a non stationary set of values into a stationary set of values by using suitable differencing.

Let $\{Z_t\}$ be a time series of non stationarity model possessing constant variance and 'd' be the quantity of differencing desired to translate $\{Z_t\}$ into a stationary set of values. If $W_t = \nabla^d Z_t$, then W_t is stationary.

The ARMA model for W_t is

$$\emptyset(B)\, W_t = \theta(B)a_t \qquad (3.15)$$

$$\emptyset(B)\nabla^d Z_t = \theta(B)a_t \quad (3.16)$$

$$g(B)\, Z_t = \theta(B)a_t \quad (3.17)$$

Where

$$g(B) = \emptyset(B)\nabla^d = \left(1 - \emptyset_1 B - \emptyset_2 B^2 - \cdots - \emptyset_p B^p\right)(1 - B)^d$$

3.15 give a time series pattern of non stationary time series, so it is called the Autoregressive Integrated Moving Average (ARIMA) process. It is represented by $Z_t \sim ARIMA(p, d, q)$.

Here the degree of AR operator is p, the degree of MA operator is q.

'd' specifies the number of differencings require to translate non stationary data to stationary data.

AR(p) implied by ARIMA (p,0,0), MA(q) implied by ARIMA (0,0,q), ARIMA(p,0,q) implies ARMA(p, q) model, that d=0 indicates stationarity of the time series.

3.2.8 Seasonal ARIMA (SARIMA) Process

The ARMA procedure is used for stationary time series, ARIMA procedure is used for non stationary time series that is non seasonal. Many time series data shows a seasonal pattern. For example, airline passengers will be more in a few months, and they will be less in other months, similar behaviour can be seen every year.

Thus we observe that a data set reveals recurrent performance when resemblances in the data set take place with same time intervals. In such cases, the series is said to be deal with seasonality. Box-Jenkins proposed a model to deal with seasonality that is Seasonal ARIMA (SARIMA) procedure. This procedure uses seasonal differencing to take out the effect of seasonality. If there are 's' seasons in a year, the seasonal differencing is given as$1 - B^s$. For time seriesZ_t,

$$W_t = (1 - B^s)\, Z_t = Z_t - Z_{t-s} \quad (3.18)$$

For monthly data $s = 12$ and $s = 4$. for data of quarters

This process is usually written as $\mathrm{SARIMA}\ (\mathrm{p, d, q})(\mathrm{P, D, Q})^S$.

The mathematical expression for $\mathrm{SARIMA}(\mathrm{p, d, q})(\mathrm{P, D, Q})^S$ is

$$\emptyset(B)\,\phi(B^s)\,(1-\mathrm{B})^d\,(1-B^s)^D Z_t = \theta(B)\,\Theta(B^s)\,a_t \qquad (3.19)$$

Where

$\emptyset(B) = 1 - \emptyset_1 B - \emptyset_2 B^2 - \cdots - \emptyset_p B^p$ is AR operator, $\phi(B^s) = 1 - \phi_1 B^s - \phi_2 B^{2s} - \cdots - \phi_P B^{Ps}$ is a seasonal AR operator, $\theta(B) = 1 - \theta_1 B - \theta_2 B^2 - \cdots - \theta_q B^q$ is MA operator and $\Theta(B^s) = 1 - \Theta_1 B^s - \Theta_2 B^{2s} - \cdots - \Theta_Q B^{Qs}$ is a seasonal MA operator. D id the number of seasonal differences required to remove the seasonal effects.

The fitting of these models for air traffic datasets is discussed at the end of the chapter.

3.2.9 Box-Jenkins Methodology

G Box and G Jenkins have given an iterative perspective to select an ARIMA technique that suits the time series data adequately. By following this methodology, one can obtain the best ARIMA model; it as well assures the parsimony principle.

The methodology consists of four steps.

i. Identify the appropriate time series model using ACF and PACF plots.
ii. Obtain the parameter estimates to the identified process by employing the least squares (LS) or Maximum Likelihood (ML) method.
iii. Test for the efficacy of the process, if the process is efficient to go to the next step otherwise, go to step 1 and review the model.
iv. Forecast the future values using the Minimum Mean Square Error (MMSE) method.

The flow chart for the Box-Jenkins methodology is given below.

Flow chart of Box-Jenkins Methodology

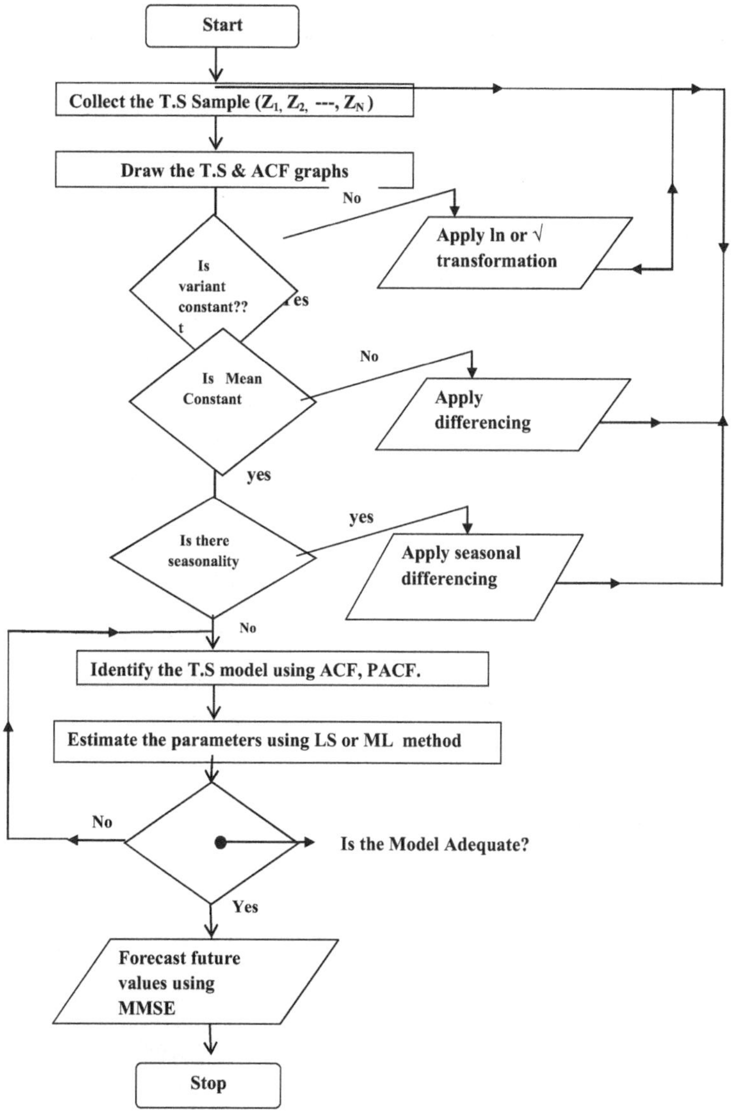

3.3 HOLT-WINTERS EXPONENTIAL SMOOTHING METHODS

The approach of Holt-Winters (HW) originated from the 1950's. As the famous experimenters had received a request from Planning and control of Industrial Operations, the Naval Research office US. They were asked to develop a low cost, high accuracy forecasting model (Holt CC, 2004). Holt created a new model that could control additive and multiplicative trend and seasonality in both the cases and standard errors. This proposed novel practice by Holt required fewer data storage, it was quick and easy to program and used robust parameters and simple basic conditions and permitted automatic adaptation. A degree scholar, Winters, tested the approach in different data sets (Holt CC, 2004). Winters found these prediction formulas were accurately, and then in the 1960s, he published his results. From then, this new approach was termed Holt-Winters (Hyndnan et al., 2008) model. The applications Holt and Holt-Winters models were rapidly spread into the software system to analyze and forecast, and they have been used even now in researches

3.3.1 Procedure of Linear Exponential Smoothing by Holt's

This procedure is further called a Double Exponential smoothing method. This method is applicable to non seasonal data, which presents the trend. This method estimates the ongoing level as well as an ongoing trend. In this, trend b_t is defined at time t as the deviation of the present and preceding level.

This method follows the expressions for level and trend as:

Level $l_t = \alpha y_t + (1 - \alpha)(l_{t-1} + b_{t-1})$ (3.20)

Trend $b_t = \beta(l_t - l_{t-1}) + (1 - \beta)b_{t-1}$ (3.21)

For $F_{t+1} = l_t + b_t$ (3.22)

In these equations, there are two positive parameters for smoothing α and β, and they are below one, i.e. $0 \leq \alpha \leq 1$ and $0 \leq \beta \leq 1$.

3.3.2. Holt-Winters Exponential Smoothing Method

The Holt linear exponential smoothing technique deals with solely level and trend but no seasonality. If a time series owns strong seasonal variation, it un clears the cycles. But studying about these seasonal patterns is very important for understanding the process under study. Holt-Winters smoothing method removes seasonality and leaves that time series with clear long time fluctuations. This method is generally called the method of Holt-Winters or Seasonal Exponential Smoothing technique or Winters procedure. It is the generalization of the Double Exponential Smoothing technique, and it deals with Level, Trend and Seasonality. This method considers α, β and γ as three parameters.

Holt-Winters techniques are classified into two methods based on the modelling of seasonality. They are the multiplicative approach and an additive approach. If the data set has approximately constant seasonal fluctuations, the additive approach is preferable otherwise, if the dataset has seasonal fluctuations deviating in proportion with the level of the dataset, the multiplicative approach is preferable. In the additive approach, the seasonal factor is presented in absolute values of the actual dataset with the same scale, and to get the level, seasonal adjustments are done to the dataset by taking the difference of the seasonal factor.

The seasonal factor sums up to roughly zero within every year. In the multiplicative technique, the seasonal factor is presented as percentages; the seasonal adjustments of the dataset are obtained by dividing the entire dataset with seasonal factor. The seasonal factor will add to roughly m within every year.

3.3.3 Holt-Winters' Additive approach

The Holt-Winters technique incorporates an expression for forecasting and three expressions for smoothing, l_t as the level, b_t as the trend, and s_t as the seasonal factor expressions having α, β , and γ as parameters for smoothing. m is utilized to characterize the seasonal frequency, which indicates total seasons per year. For instance, $m = 4$ for quarterly data, and for monthly data $m = 12$

In the additive approach, the component structure is

Level $l_t = \alpha(y_t - s_{t-m}) + (1 - \alpha)(l_{t-1} + b_{t-1})$ (3.23)

Trend $\quad b_t = \beta(l_t - l_{t-1}) + (1 - \beta)b_{t-1}$ (3.24)

Seasonal $\; s_t = \gamma(y_t - l_{t-1} - b_{t-1}) + (1 - \gamma)s_{t-m}$ (3.25)

Forecast $\hat{y}_{t+h/t} = l_t + hb_t + s_{t+h-m(k+1)}$ (3.26)

Where m is the span of the seasonal rhythm, $0 \leq \alpha \leq 1$, $0 \leq \beta \leq 1$, and $0 \leq \gamma \leq 1$. Seasonal expression locates deviations of the ongoing level and the value in that time of the seasonal rotation. After that, it is merged to a forecast at a similar spike in the rotation.

Here k is the truncation of (h-1)/m that makes clear that the seasonal factor estimates applied for forecasting belong to the last year of the dataset. The expression of level can be obtained by taking a weighted mean of the values adjusted seasonally $(y_t - s_{t-m})$ as well as the non seasonal forecast $(l_{t-1} + b_{t-1})$ at time t.

The expression for the trend is the same as Holt's linear expression. The expression for seasonality is a weighted mean of present seasonal record, $(y_t - l_{t-1} - b_{t-1})$ and the seasonal record of the parallel season previous year that is m time periods earlier.

3.3.4 Holt-Winters Multiplicative technique

The multiplicative technique of Holt-Winters incorporates an expression for forecasting and three expressions for smoothing, l_t as the level, b_t as the trend, and s_t as the seasonal factor expressions having α, β andγ as parameters for smoothing. m is utilized to characterize the seasonal frequency, which indicates total seasons per year. For instance, $m = 4$ for quarterly data, and for monthly data $m = 12$

In the multiplicative approach, the component structure is

Level $\; l_t = \alpha\left(\frac{y_t}{s_{t-m}}\right) + (1 - \alpha)(l_{t-1} + b_{t-1})$ (3.27)

Trend $\quad b_t = \beta(l_t - l_{t-1}) + (1 - \beta)b_{t-1}$ (3.28)

Seasonal $\; s_t = \gamma\left(\frac{y_t}{l_{t-1} - b_{t-1}}\right) + (1 - \gamma)s_{t-m}$ (3.29)

Forecast $\quad \hat{y}_{t+h/t} = (l_t + hb_t)\, s_{t+h-m(k+1)}$

Where m is the span of the seasonal rhythm, $0 \leq \alpha \leq 1$, $0 \leq \beta \leq 1$, and $0 \leq \gamma \leq 1$. Seasonal expression locates the ratio of the ongoing level and the value in that time of the seasonal rotation. After that, it is multiplied to a forecast at a similar spike in the rotation.

Here k is the truncation of (h-1)/m that makes clear that the seasonal factor estimates applied for forecasting belong to the last year of the dataset. The expression of level can be obtained by taking a weighted mean of the values adjusted seasonally $(\frac{y_t}{s_{t-m}})$ as well as the non seasonal forecast $(l_{t-1} + b_{t-1})$ at time t.

The expression for the trend is the same as Holt's linear expression. The expression for seasonality is a weighted mean of present seasonal record $\left(\frac{y_t}{l_{t-1}-b_{t-1}}\right)$ and the seasonal record of the parallel season previous year that is m time periods earlier.

The components of the Holt-Winters model can be determined by finding the evaluates of the constants α, β and γ to predict starting values. Then these values are used in finding the exact evaluates of the parameter that reduces the error sum of squares of the next step forecast. The level l_t is calculated by combining the actual and the adjusted at time t, eliminates the effect of seasonality in the dataset. The determination of trend b_t is taking consecutive deviation of two successive levels. Like the Holt model, the seasonal expression s_t can be found by merging the actual and seasonal adjusted record at time t (Souza et al., 2008).

3.4 MATERIALS AND METHODS

In previous, we have discussed Stochastic models and techniques of Exponential smoothing for forecasting the time series. Here these procedures have been applied to real data sets to forecast. In this thesis, time series data considered is the number of passengers travelled monthly by Air India and Indigo scheduled Domestic services. The number of passengers is considered as air traffic. The data has been taken from the official website of DGCA (Directorate General of Civil Aviation). Results were carried out using SPSS software.

The best model among different models, namely AR, MA, ARMA, ARIMA, SARIMA and Holt Winters' procedures discussed above, has been identified using performance evaluation metrics MAE, MSE, RMSE, MAPE, BIC and R Square. The results obtained by various models have been presented in the form of tables. Plots of actual and forecasted values also presented in this chapter to observe the adequacy of various models.

3.5 AIR INDIA AIR TRAFFIC DATA

The subject of the analysis is Air India, with the number of passengers travelled by scheduled domestic flights being the focus. The research period covers the years from 2012 to 2018. Data were presented in monthly cycles. The data was considered from 2012 January to 2018 December.

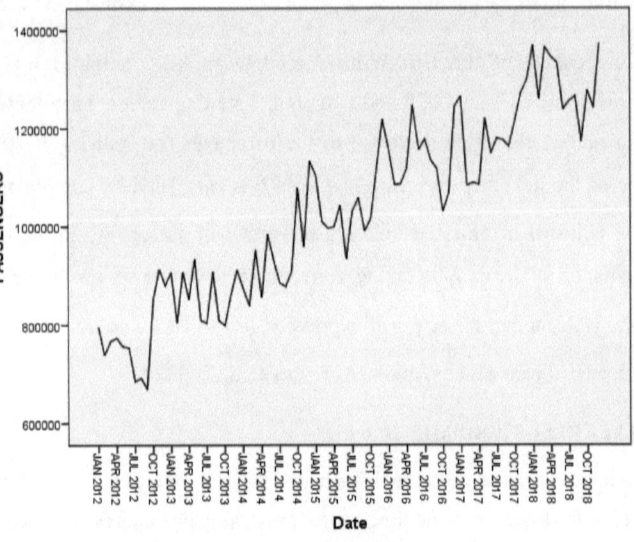

Figure 3.1. Air India air traffic data (Jan 2012 to Dec 2018)

Figure 3.1 represents the sequence plot of air traffic data regarding the number of air passengers travelled by Air India on scheduled domestic services. The basic statistical characteristics of Air India air traffic data is presented in table 3.1.

Statistic	Value
Minimum	669342
Maximum	1372904
Mean	1032337
Standard deviation	186066.2
Skewness	-0.09117512
Kurtosis	1.923432

Table 3.1.Statistical characteristics of Air India Air traffic

3.5.1 ARIMA Model

Stationarity of the data set is checked by using KPSS test. The results are given below.

Eta(Observed value)	2.697
Eta(Critical value)	0.449
p-value(one-tailed)	<0.0001
Alpha	0.05

Table 3.1.a KPSS test results for Air India Air traffic

Test interpretation:

H_0 : The series is stationary.

H_1 : The series is not stationary.

As the computed p-value is lower than the significance level alpha=0.05, we should reject the null hypothesis H_0, and accept the alternative hypothesis H_a.

Figure 3.1 also supports that the air traffic data is non stationary. To achieve stationarity in mean the series has been differenced once, to eliminate non stationarity in variation, the data set was reconstructed by employing the Natural log transformation. Now again we applied KPSS test for conforming the stationarity.

Eta(Observed value)	0.026
Eta(Critical value)	0.442
p-value(one-tailed)	0.997
Alpha	0.05

Table 3.1.b. KPSS test results for Air India Air traffic(transformed series)

Test interpretation:

H_0 : The series is stationary.

H_1 : The series is not stationary.

As the computed p-value is greater than the significance level alpha=0.05, we accept the null hypothesis H_0. i.e., the series is stationary.

Figure 3.2 represents the stationary air traffic data.

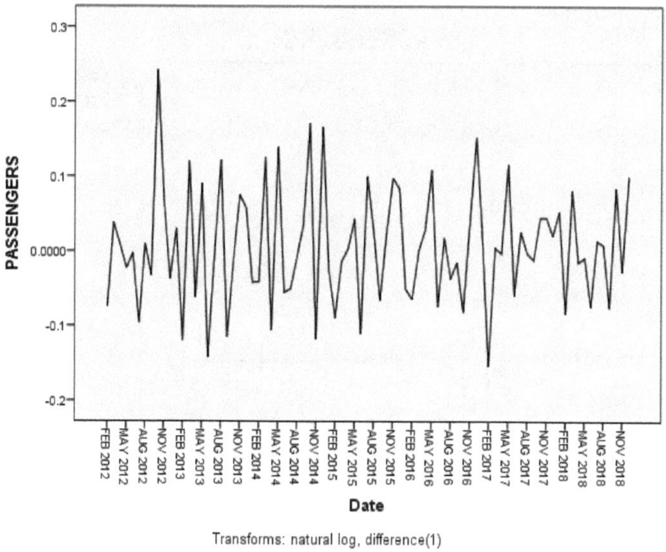

Figure 3.2 Stationary air traffic data (Air India)

The sub class of ARIMA structure can be recognized using PACF and ACF plots. From the ACF plot Moving Average (MA) component will be identified, and from the PACF plot Auto Regression (AR) component will be identified. Figure 3.3, as well as 3.4, represents the plots of ACF and PACF, respectively.

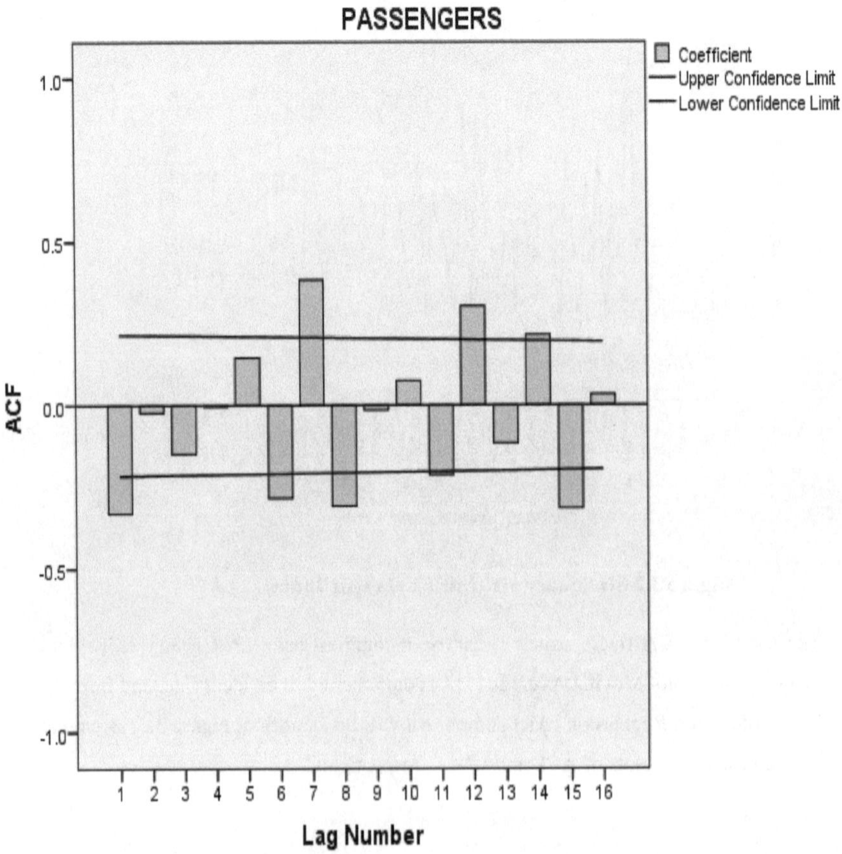

Figure 3.3.Chart of ACF of Air India Air Traffic data

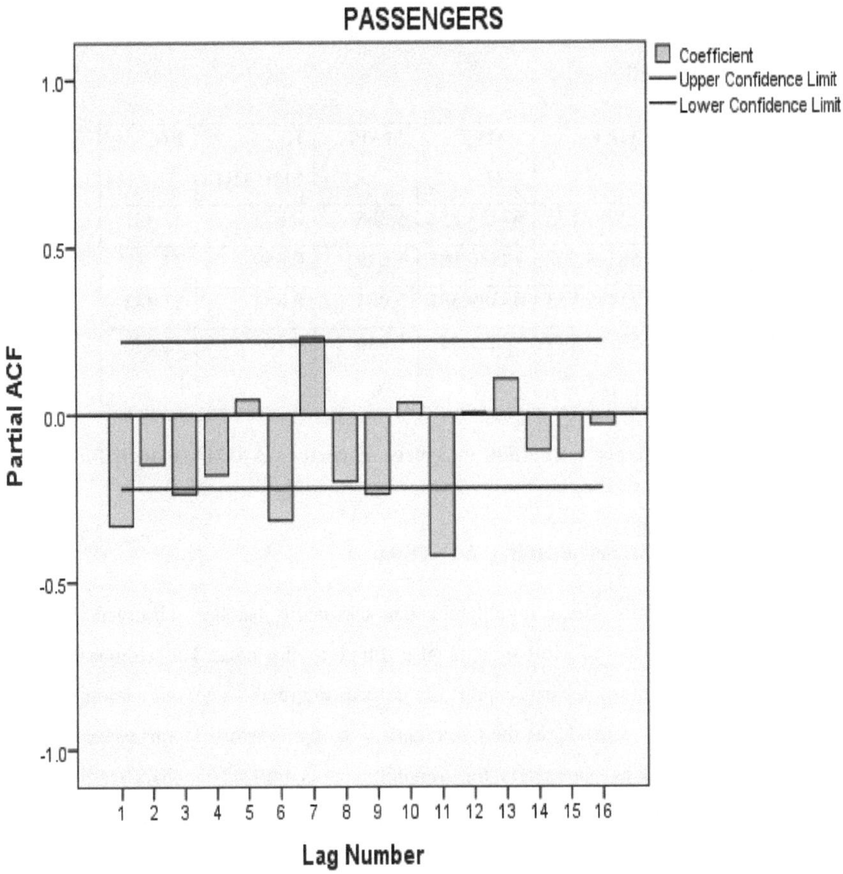

Figure 3.4.Representation of PACF of Air India Air Traffic data

The plots of PACF and ACF bestow the possible ARIMA models as

- ARIMA (0, 1, 1)
- ARIMA (1, 1, 0)
- ARIMA (1, 1, 1)
- SARIMA (0, 1, 0)(1, 1, 0)

The performance of these models has been checked by using R Square, MSE, RMSE, MAE, MAPE and BIC values. These values for different ARIMA structures are furnished in table 3.2

MODEL	MAE	RMSE	MAPE	R SQUARE	BIC
ARIMA(0,1,1)	53997.835	67428.851	5.368	0.873	22.451
ARIMA(1,1,0)	56344.568	70836.185	5.579	0.859	22.549
ARIMA(1,1,1)	**51419.953**	**65389.581**	**5.084**	**0.882**	**22.442**
SARIMA(0,1,0)(1,1,0)	61700.147	78229.883	5.980	0.767	22.595

Table 3.2 Performance evaluation measures of various ARIMA models for Air India

3.5.2 Additive Approach of HOLT WINTERS

In Air India air traffic data, a roughly constant seasonality has been observed. Thus the additive approach of Holt-Winters also fitted to the data. The Holt-Winters technique incorporates an expression for forecasting and three expressions for smoothing, l_t as the level, b_t as the trend, and s_t as the seasonal factor expressions having α, β, and γ as parameters for smoothing. m is utilized to characterize the seasonal frequency, which indicates total seasons per year. m=12 in the present study.

The model parameters are presented in table 3.3.

		Parameter	Estimate
HOLT-WINTERS' Additive model	Level	Alpha(α)	0.400
	Trend	Beta(β)	0.000003758
	Season	Gamma(γ)	0.00005694

Table 3.3 Parameter estimates of Winters' additive method of Air India

Performance evaluation measures are presented in table 3.4

Measure	Value
MAE	41178.488
RMSE	51298.805
MAPE	4.116
R SQUARE	0.926
BIC	21.849

Table 3.4 Performance evaluation measures of Winters' additive method of Air India

3.5.3 Numerical Results

In practice, the low value of MAE, RMSE, MAPE and BIC indicate the best fit for the model, and the greater R^2 value indicates an exact prediction. Also, less than 10% value of MAPE refers to an extremely accurate forecast; fewer than 20% refer to possibly satisfactory and exceeding 30% indicates probably unreliable. From table 3.2 it can be observed that MAPE values of all the models are less than 10%, all the models are good for forecasting but MAE, RMSE, MAPE and BIC values of ARIMA (1, 1, 1) are less than ARIMA (0, 1, 1), ARIMA(1, 1, 0) and SARIMA (0, 1, 0) (1, 1, 0) and also R square value of ARIMA (1, 1, 1) is greater than ARIMA (0, 1, 1), ARIMA (1, 1, 0) and SARIMA (0, 1, 0)(1, 1, 0). Hence ARIMA (1, 1, 1) is optimum for this air traffic and observed and fitted plots of these models have been presented in Figures 3.5, 3.6, 3.7 and 3.8.

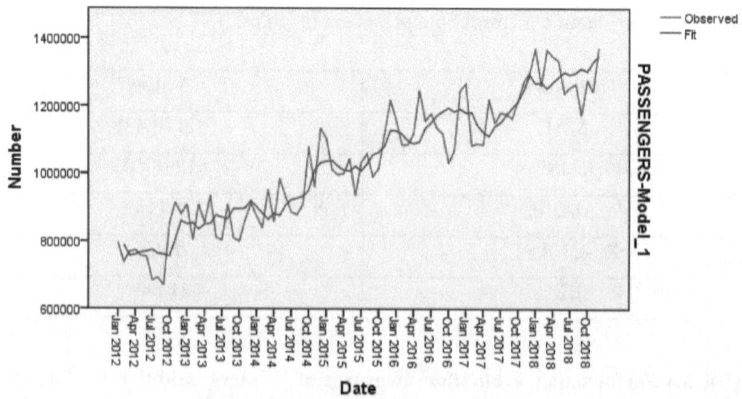

Figure 3.5 Representation of Original and Fitted points by ARIMA (0,1,1) of Air India

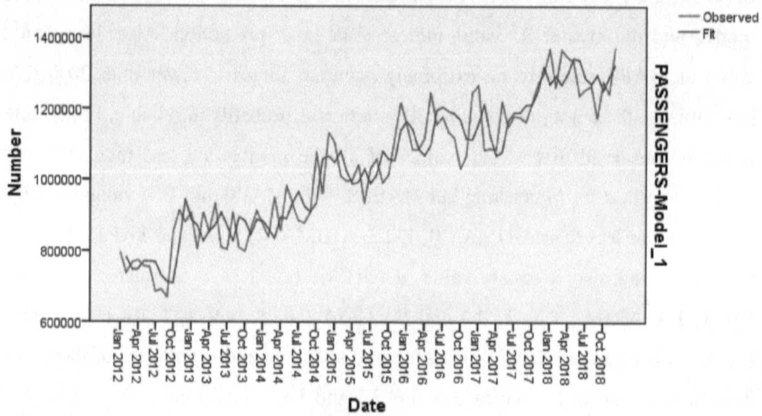

Figure 3.6 Representation of Original and Fitted points by ARIMA (1,1,0) of Air India

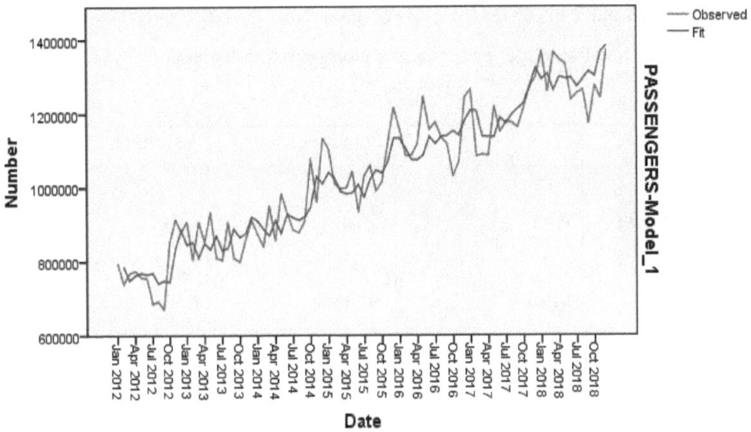

Figure 3.7 Graph of Original and Fitted points by ARIMA (1, 1, 1) of Air India

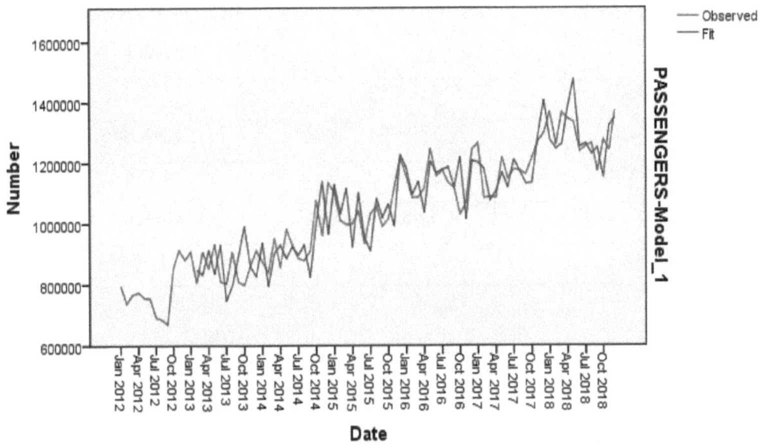

Figure 3.8 Representation of Original and Fitted points by SARIMA(0,1,0)(1,1,0) of Air India

From all the above figures, it can be observed that the observed and fitted values are closer to each other in ARIMA (1,1,1) (Figure 3.7). This also indicates the adequacy of the model, and the model parameters are presented in table 3.5.

		Parameter	**Estimate**
ARIMA(1,1,1)	AR1	Ø	0.347
	MA1	Θ	0.999

Table 3.5 Model parameters' estimates of ARIMA (1,1,1) of Air India

The adequacy of **the Additive** approach of **Holt-Winters** and **ARIMA** is checked by the MAE (Mean Absolute Error), RMSE (Root Mean square Error), MAPE (Mean Absolute Percentage Error), R^2 and BIC (Bayesian Information Criterion).

MEASURE	**ARIMA(1,1,1)**	**HOLT- WINTERS' ADDITIVE**
MAE	51419.953	**41178.488**
RMSE	65389.581	**51298.805**
MAPE	5.084	**4.116**
R SQUARE	0.882	**0.926**
BIC	22.442	**21.849**

Table 3.6 Performance evaluation measures of ARIMA and Holt-Winters' models for Air India

For this data, both the models Holt-Winters (HW) Additive and ARIMA have MAPE values less than 10%, which shows both are potentially very good for forecasting, but the Additive technique of Holt-Winters' MAPE is 4.116 also of ARIMA is 5.084. MAE of HW Additive technique is 41178.488 also of ARIMA is 51419.953. RMSE of HW Additive technique is 51298.805 also of ARIMA is 65389.581. BIC of HW Additive technique is 21.849 also of ARIMA is 22.442. R-squared value of HW Additive technique is 0.926 also of ARIMA is 0.882. Thus MAE, RMSE, MAPE and

BIC are less for Holt-Winters' Additive model than ARIMA. Also, the R-squared value is high for the Holt-Winters Additive model than ARIMA, which shows the Additive method of Holt-Winters is superior to ARIMA for this data.

The plot of observed and fitted values by Holt-Winters' additive model has been presented in 3.9, and the representation of original and forecasted points by the model has been presented in 3.10.

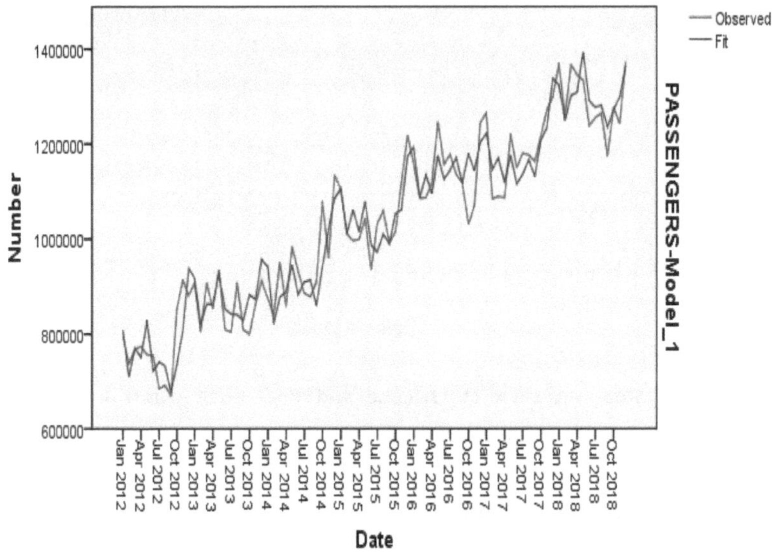

Figure 3.9 Representation of Original and Fitted points by Winters' additive method of Air India

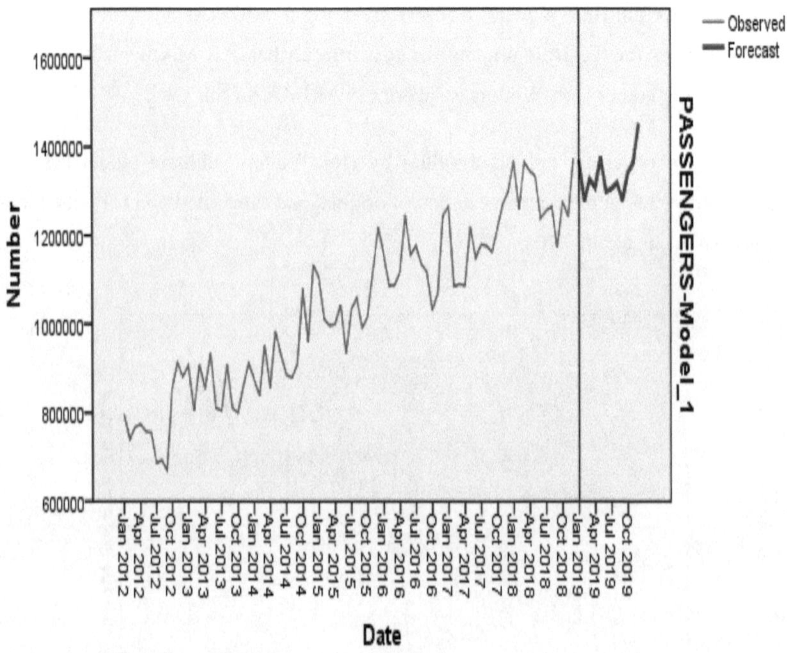

Figure 3.10 Representation of Original and Forecasted points by Winters' additive method of Air India

3.6 INDIGO AIR TRAFFIC DATA

The subject of the analysis is Indigo, with the number of passengers travelled by scheduled domestic flights being the focus. The research period covers the years from 2012 to 2018. Data were presented in monthly cycles. The data was considered from January 2012 to December 2018.

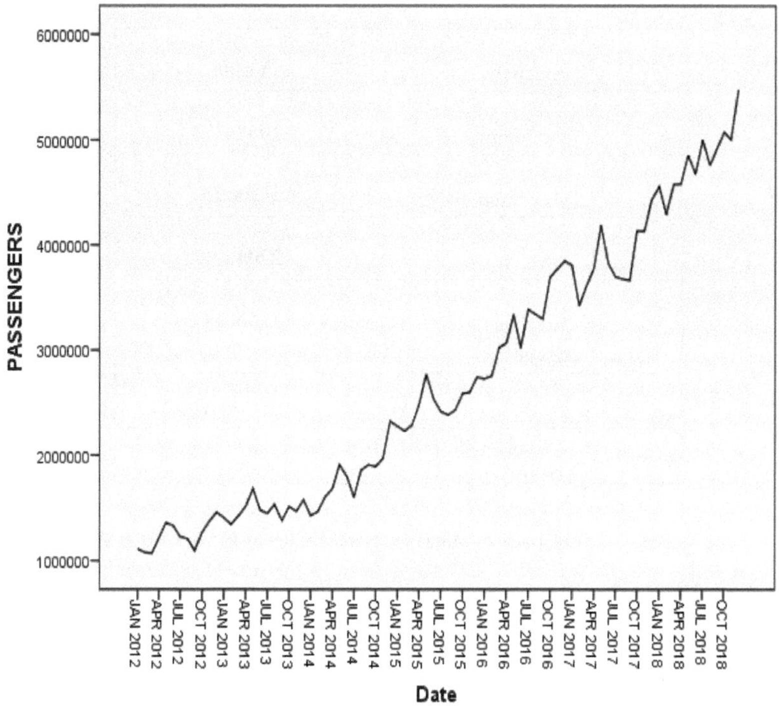

Figure 3.11 Indigo air traffic data (Jan 2012 to Dec 2018)

Figure 3.11 represents the sequence plot of air traffic data regarding the number of air passengers travelled by Indigo on scheduled domestic services. The basic statistical characteristics of Indigo air traffic data is presented in table 3.7.

Statistic	Value
Minimum	1068637
Maximum	5478523
Mean	2703594
Standard deviation	1266051
Skewness	0.58700
Kurtosis	1.86014

Table 3.7.Statistical characteristics of Indigo Air traffic

3.6.1 ARIMA Model

Stationarity of the data set is checked by using KPSS test. The results are given below.

Eta(Observed value)	2.819
Eta(Critical value)	0.449
p-value(one-tailed)	< 0.0001
Alpha	0.05

Table 3.7.a. KPSS test results for Indigo Air traffic

Test interpretation:

H_0 : The series is stationary.

H_1 : The series is not stationary.

As the computed p-value is lower than the significance level alpha=0.05, we should reject the null hypothesis H_0, and accept the alternative hypothesis H_a.

Figure 3.11 also supports that the air traffic data is non stationary. To achieve stationarity in mean the series has been differenced once, to eliminate non stationarity in variation, the data set was reconstructed by employing the Natural log transformation. Now again we applied KPSS test for conforming the stationarity.

Eta(Observed value)	0.249
Eta(Critical value)	0.442
p-value(one-tailed)	0.195
Alpha	0.05

Table 3.7.b. KPSS test results for Indigo Air traffic(transformed series)

Test interpretation:

H_0 : The series is stationary.

H_1 : The series is not stationary.

As the computed p-value is greater than the significance level alpha=0.05, we accept the null hypothesis H_0. i.e., the series is stationary.

Figure 3.12 represents the stationary air traffic data.

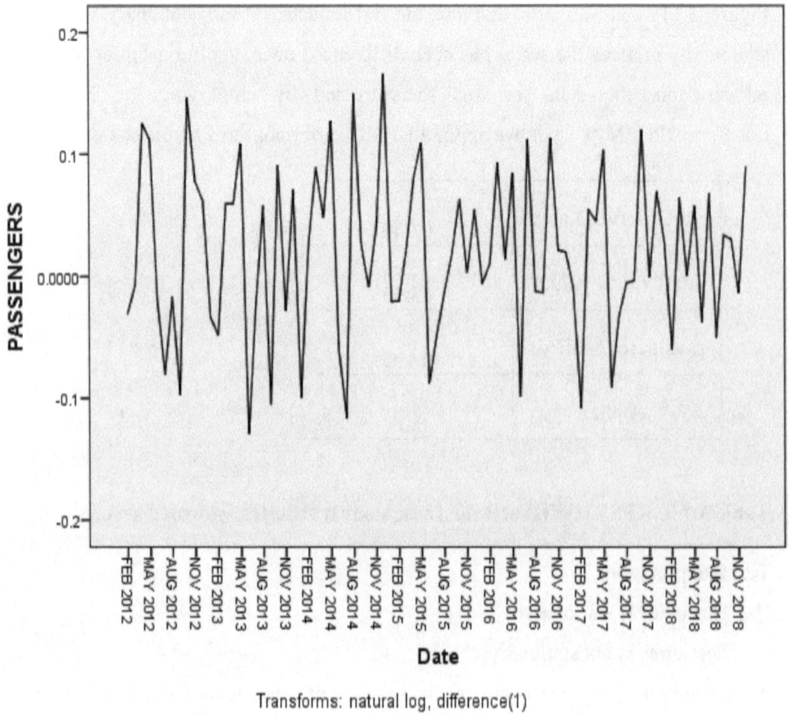

Figure 3.12 Stationary air traffic data (Indigo)

The sub class of ARIMA structure can be recognized using PACF and ACF plots. From the ACF plot Moving Average (MA) component will be identified, and from the PACF plot Auto Regression (AR) component will be identified. Figure 3.13 and figure 3.14 represents the plots of ACF and PACF, respectively.

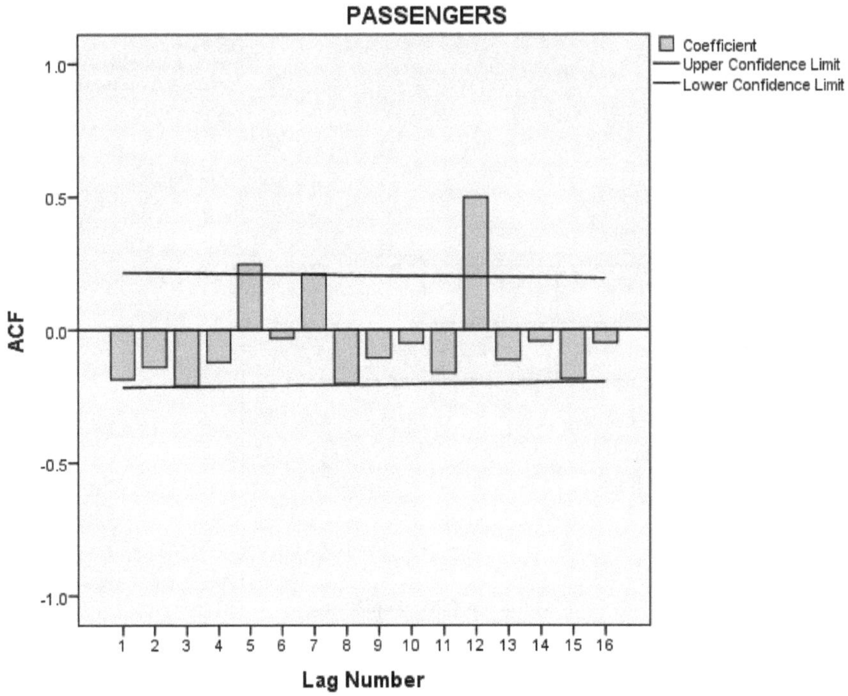

Figure 3.13 Chart of ACF of Indigo Air Traffic data

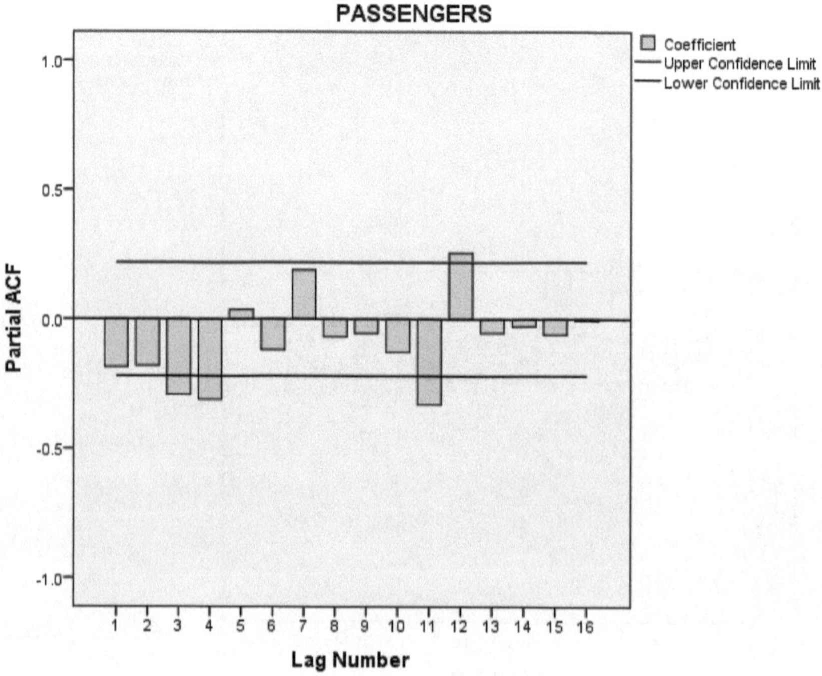

Figure 3.14 Chart of PACF of Indigo Air Traffic data

The plots of PACF and ACF bestow the possible ARIMA models as

- ARIMA (0, 1, 0)
- SARIMA (0, 1, 0)(0, 1, 1)

The performance of these models has been checked by using R Square, MSE, RMSE, MAE, MAPE and BIC values. These values for different ARIMA structures are placed in table 3.8

MODEL	MAE	RMSE	MAPE	R SQUARE	BIC
ARIMA(0,1,0)	143238.405	181805.943	5.680	0.980	24.381
SARIMA(0,1,0) (0,1,1)	**116813.034**	**149450.095**	**4.228**	**0.984**	**23.889**

Table 3.8 Performance evaluation measures of various ARIMA models

3.6.2 Additive Approach of HOLT-WINTERS

In Indigo air traffic data, a roughly constant seasonality has been observed. Thus the additive approach of Holt-Winters also fitted to the data. The Holt-Winters technique incorporates an expression for forecasting and three expressions for smoothing, l_t as the level, b_t as the trend, and s_t as the seasonal factor expressions having α, β, and γ as parameters for smoothing. m is utilized to characterize the seasonal frequency, which indicates total seasons per year. m=12 in the present study.

The model parameters are presented in table 3. 9.

		Parameter	Estimate
HOLT-WINTERS' Additive model	Level	α	0.699
	Trend	β	0.0000004633
	Season	γ	0.000008526

Table 3.9 Parameter estimates of Winters' additive method of Indigo

Performance evaluation measures are presented in table 3.10.

Measure	Value
MAE	92660.607
RMSE	118966.697
MAPE	4.010
R SQUARE	0.991
BIC	23.531

Table 3.10 Performance evaluation measures of Winters' additive method of Indigo

3.6.3 Numerical Results

From table 3.8 it can be observed that MAPE values of all the models are less than 10%, all the models are good for forecasting but MAE, RMSE, MAPE and BIC values of SARIMA (0, 1, 0) (0, 1, 1) are less than ARIMA(0,1,0) and also R square value of SARIMA(0,1,0)(0,1,1) is greater than ARIMA(0,1,0). Hence SARIMA (0,1,0) (0,1,1) structure is optimum for this data, and observed and fitted plots of these models have been presented in Figures 3.15, 3.16.

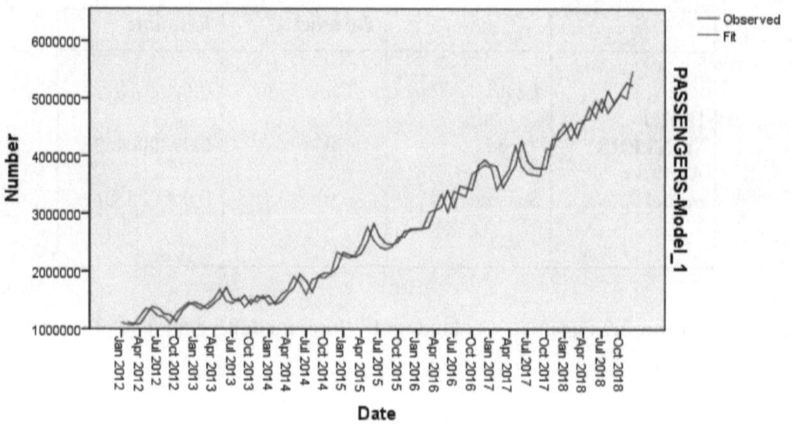

Figure 3.15 Representation of Original and Fitted points by ARIMA (0,1,0) of Indigo

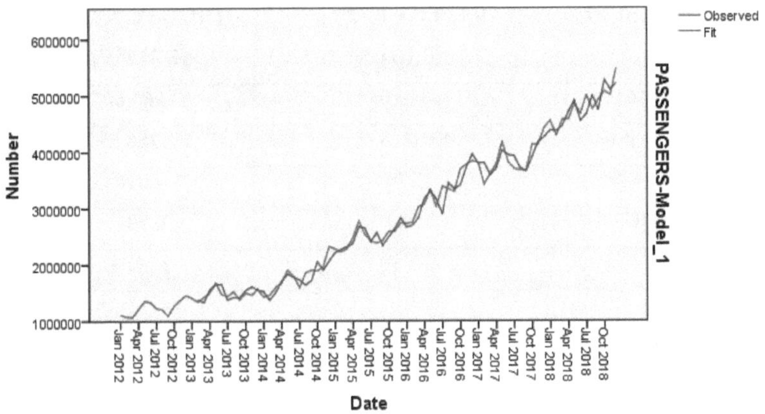

Figure 3.16 Representation of Original and Fitted points by SARIMA (0,1,0)(0,1,1) of Indigo

From all the above figures, it can be observed that the observed and fitted values are closer to each other in SARIMA (0,1,0) (0,1,1) (Figure 3.16). This also indicates the adequacy of the model, and the model parameters are presented in table 3.11.

		Estimate
SARIMA(0,1,0)(0,1,1)	Difference	1
	Seasonal Difference	1
	MA, Seasonal lag1	0.523

Table 3.11 Estimates of SARIMA (0,1,0)(0,1,1) of Indigo

The adequacy of **the Additive** approach of **Holt-Winters** and **SARIMA** are checked by the MAE (Mean Absolute Error), RMSE (Root Mean square Error), MAPE (Mean Absolute Percentage Error), R^2 and BIC (Bayesian Information Criterion).

MEASURE	SARIMA(0,1,0)(0,1,1)	HOLT- WINTERS ADDITIVE
MAE	116813.034	**92660.607**
RMSE	149450.095	**118966.697**
MAPE	4.228	**4.010**
R SQUARE	0.984	**0.991**
BIC	23.889	**23.531**

Table 3.12 Performance evaluation measures of SARIMA and Holt-Winters' models for Indigo

For this data, both the models Holt-Winters (HW) Additive and SARIMA have MAPE values less than 10%, which shows both are potentially very good for forecasting, but the Additive approach of Holt-Winters' MAPE is 4.010 and of SARIMA is 4.228. MAE of HW Additive technique is 92660.607 and of SARIMA is 116813.034. RMSE of HW Additive technique is 118966.697 and of SARIMA is 149450.095. BIC of HW Additive technique is 23.531 and of SARIMA is 23.889. R-squared value of HW Additive technique is 0.991 and of SARIMA is 0.984. Thus MAE, RMSE, MAPE and BIC are less for Holt-Winters' Additive model than SARIMA. Also, the R-squared value is high for Holt-Winters' Additive model than SARIMA, which shows the Additive method of Holt-Winters is superior to SARIMA for this data.

The plot of observed and fitted values by Holt-Winters' additive model has been presented in 3.17, and the representation of original and forecasted points by the model has been presented in 3.18.

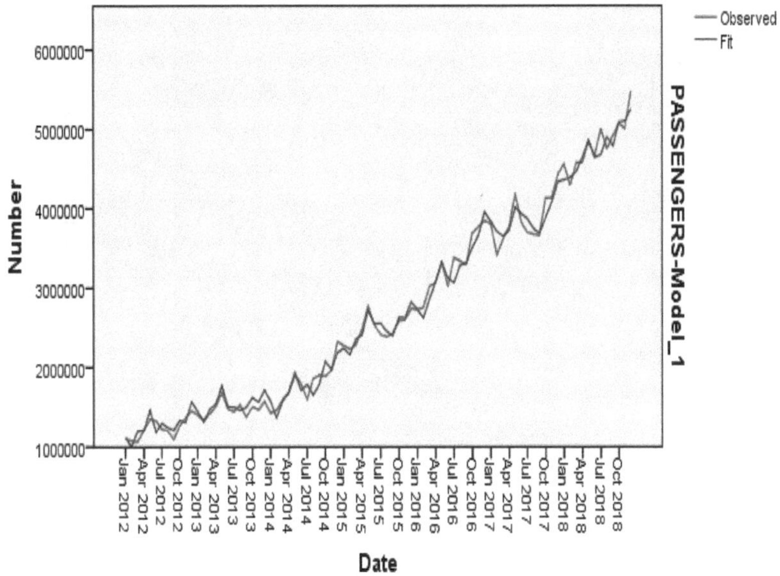

Figure 3.17 Representation of Original and Fitted points by Winters' additive method of Indigo

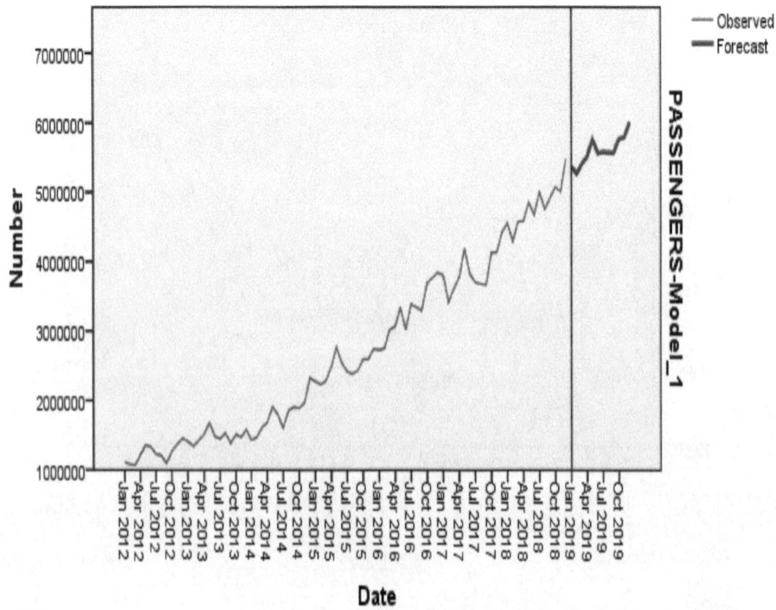

Figure 3.18 Representation of Original and Forecasted points by Winters' additive method of Indigo

3.7 CONCLUSION

In this chapter, we have considered two real datasets regarding the number of passengers travelled monthly by Air India and Indigo Scheduled Domestic services. Here the number of passengers treated as air traffic is time series data. Various Box-Jenkins approaches and Exponential smoothing models have been fitted to these air traffic datasets.

For Air India air traffic data Holt-Winters Additive and ARIMA (1,1,1) have been identified as best by using performance evaluation measures. Both the models are extremely very good for forecasting as their MAPE values are less than 10%. The Holt-Winters Additive model has been identified as the best model for Air India air traffic data by comparing the performance evaluation measures.

For Indigo air traffic data, the Additive technique of Holt-Winters and SARIMA (0,1,0) (0,1,1) have been identified as best by using performance evaluation measures. Both the models are extremely very good for forecasting as their MAPE values are less than 10%. By comparing the performance evaluation measures, the Holt-Winters Additive model has been identified as the best model for Indigo air traffic data. This kind of analysis is useful for air service providers to design strategies for the development of services.

CHAPTER-IV

AIR TRAFFIC FORECASTING USING ARTIFICIAL NEURAL NETWORKS

4.1 INTRODUCTION

In the former section, diverse stochastic models and exponential smoothing models have been discussed to analyze and forecast air traffic. In recent years Artificial Neural Networks (ANNs) have become more popular as a revolutionary approach to forecast the time series. The present chapter discusses Artificial Neural Networks to forecast air traffic.

The field of Neural Network was established before the invention of computers, but Neural Network imitations are of contemporary evolution. In 1943 the elemental artificial neuron was invented by Warren Mc Culloch, a neurophysiologist. After that, artificial neurons were proposed by Hebbian (1949) and Perceptron (1958). The single layer networks, which are called perceptrons, were introduced by Frank Rosenblatt.

Neural Network (NN) solves the problem by a different approach than that of conventional computers. Neural Network exercise instruction alike as human brain exercises. The NN is made of an immense amount of interlinked neurons, which are the processing elements that work parallel in solving a particular problem.

ANN is an information processing pattern that is influenced by the manner organic neurological scheme like the spinal cord, brain exercise the information. Artificial Neural Network (ANN) has wider applications to real world problems. They are well suited for structure identification and categorization.

ANNs got greater prominence in the analysis of time series as well as forecasting. The ANN aims to pursue to identify structures in a given set. After the NN trained on a given set, it detects a similar pattern in future data to make predictions (Mohie El-Din et al., 2017)

4.2 FEED-FORWARDED NETWORKS

Feed-Forwarded ANNs enable signals to proceed from inserted data to result. In this, the result of some layer will not alter the similar layer, i.e. there is no feedback (loops). Feed-Forwarded Networks are uncomplicated architectures, and they relate inserted data to result. Feed-Forward Networks are most-widely employed for structure identification. This organization is as well termed as top-down or bottom-up.

Activation function (AF) is applied to the weighted sum of the inputs of a neuron to give the output. The activation functions that are commonly employed are:

- Identify function: $f(x) = x$
- Binary step function: $f(x)\begin{cases} = 1\, when\, x < \theta \\ = 0\, when\, x \geq \theta \end{cases}$
- Sigmoid function: $\frac{1}{1+e^{-x}}$
- Bipolar sigmoid function: $(1 - e^{-x)}(1 + e^{-x)^{-1}}$

Two types of training algorithms used to train Neural Networks.

1. In supervised learning, it checks at every step the input that is applied with the desired output supplied by the model. It takes less no of iterations, less time, and it produces more accuracy. Some of the supervised learning are Perceptron, Delta learning law and Back Propagation etc.
2. UN supervised learning, when the input is applied at every stage again, it passes through the network. The process stops when the two adjacent outputs are the same. It takes more number of iterations, more time and gives less accuracy. Examples are Hebbaian learning, Competitive Learning and self organizing maps etc.

4.3 MULTI-LAYER PERCEPTRON

Multi-Layer Perceptron (MLP) is the most widely used network structure of Artificial Neural Network in time series analysis. A Multi-Layer perceptron is a class of feed forward Artificial Neural Network.

An MLP incorporates not less than three layers of nodes: one Input layer, 1 output layer, as well as 1 or higher hidden layers (Masters T, 1983). Input layer acquires the action, output layer creates a resolution or projection on the inserted data and allying those two, a discretional amount of hidden layers which are the accurate reckoning mechanism of the Multi-Layer Perceptron.

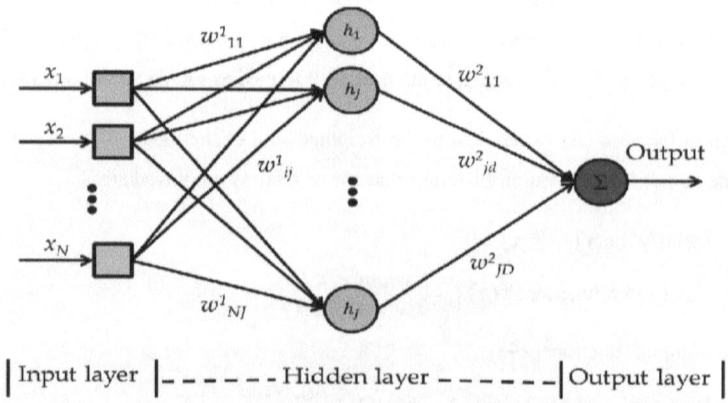

FIGURE 4.1. Picture of Multi-Layer Perceptron network.

MLP implements a supervised studying course of action called back propagation to train. Back propagation is a prominent mathematical kit to upgrade the precision of projections in machine learning and data mining. A NN cultivates the action of the inserted data further along with its parameters in the direction of the instant of the resolution, after that back propagates instruction on the error, inverse by the way of a network, with this it will revise the parameters.

It occurs in stages

- The network creates speculation on data based on the parameters.
- An error of the network is enumerated using the loss function.
- This error is back propagated to regulate the erroneous specified parameters.

4.4 BACK PROPAGATION ALGORITHM

Back propagation captures the error related to erroneous speculation via neural networks, as well as utilizes the error to regulate the parameters of neural networks in the track of minimum error. The weights for a particular node are adjusted in direct proportion to the connected units of the node. The output is produced by applying an activation function to the weighted sum of inputs of a neuron.

The activation function employed in this investigation is the sigmoid function.

Sigmoid function: $\frac{1}{1+e^{-x}}$ (4.1)

4.5 MODEL DEVELOPMENT BY FEED-FORWARDED MULTI-LAYER PERCEPTRON NETWORKS

Using the back propagation rule, the MLP learning algorithm includes initializing weights (to small random values) and transfer function and adjusting weights by **starting** from the output layer and working backwards.

$$W_{ij}\ (t+1) = W_{ij}\,(t) + \boldsymbol{\eta\delta_{pj}\,O_{pj}} \quad (4.2)$$

Where $W_{ij}\,(t)$ represents the weights from node i to node j at time t,

$\boldsymbol{\eta}$ is a gain term, and $\boldsymbol{\delta_{pj}}$ is an error term for pattern p on node j (where the sum is over the k nodes in the following layer)

For output layer units: $\boldsymbol{\delta_{pj}} = \boldsymbol{kO_{pj}(1 - O_{pj})\ (t_{pj} - O_{pj})}$ (4.3)

For hidden layer units: $\boldsymbol{\delta_{pj}} = \boldsymbol{kO_{pj}(1 - O_{pj})\ \Sigma\delta_{pk}W_{jk}}$. (4.4)

A unit in the output layer determines its activity by following a 2- step procedure.

Step1: It computes the total weighted input X_j using the formula $X_j = \sum_i \boldsymbol{y_i W_{ij}}$, Where $\boldsymbol{y_i}$ is the activity level of the jth unit in the previous layer.

Step2: Calculate the activity $\boldsymbol{y_j}$ using the sigmoid function of the total weighted input once the activities of all output units have been determined, the network computes the error E

$$y_j = \frac{1}{1+e^{-xj}} \quad (4.5)$$

$$E = \frac{1}{2}\ \boldsymbol{\Sigma(y_i - d_i}\ ^2 \quad (4.6)$$

Here y_i indicates the activity intensity of the jth element in the peak layer, and d_j indicates the expected result of the jth element.

4.6 MATERIALS AND TECHNIQUES

4.6.1 Artificial Neural Networks Architecture for AIR INDIA

In the investigation, Multi-Layer Perceptron (MLP) network has been used for the prediction of air traffic flow. For the development of the Artificial Neural Network (ANN) model, 84 months of data points regarding the number of passengers travelled by Air India domestic services have been taken. Air traffic was considered as an input variable, and one hidden layer was considered. Different Artificial Neural Network (ANN) models have been developed on the training data set. In the present study ERROR, MSE, RMSE, and MAE values have been used to evaluate the performance of the model and predicted results. The specification of all the models has been presented in table 4.1.

4.6.1.1 Numerical Results

It can be observed from table 4.1 that model 3, i.e., a neural network with 3 hidden neurons, has minimum Error, Mean Square Error (MSE), Root Mean Square Error (RMSE), Mean Absolute Error (MAE) values. Hence it is used to forecast the future air traffic.

Model	**Hidden layer**	**Hidden neurons**	**ERROR**	**MSE**	**RMSE**	**MAE**
M1	1	1	0.31842	5917114685.869600	76922.7839	64112.5062
M2	1	2	0.31776	5903147003.622080	76831.9400	63991.8400
M3	1	3	0.31696	5887281357.835080	76728.6215	63805.4018

Table 4.1 Different Neural Network models' ERROR, MSE, RMSE and MAE values of Air India

The actual and Forecasted plot by model 3 has been presented in figure 4.2, the representation of Original and Projected points has been presented in figure 4.3, Original versus Projected plot has been presented in figure 4.4 as well as Residual graph presented in 4.5.

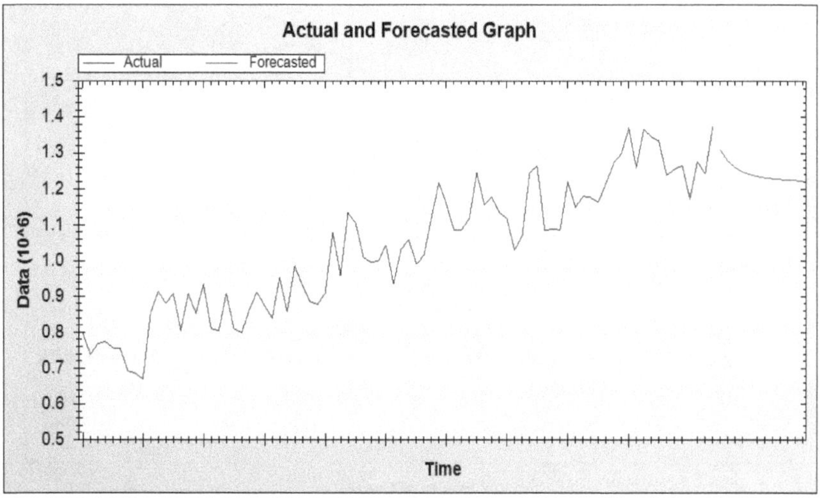

Figure 4.2 Plot of Actual and Forecasted Air India air traffic

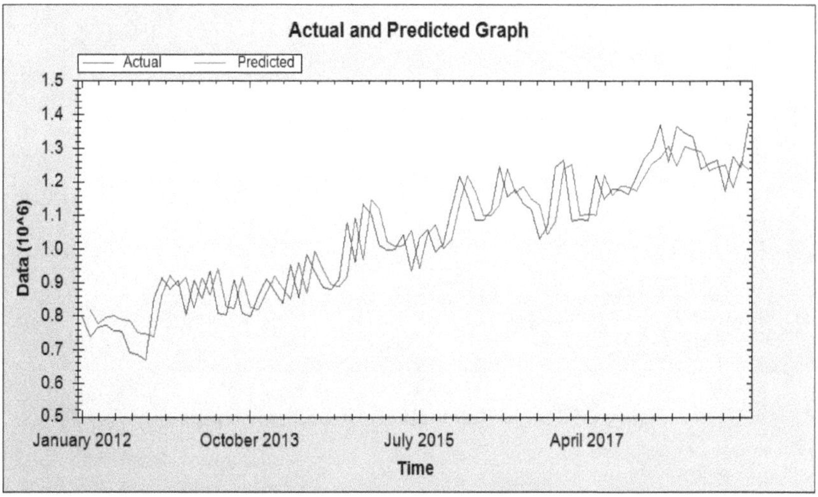

Figure 4.3 Representation of Original and Projected Air India air traffic

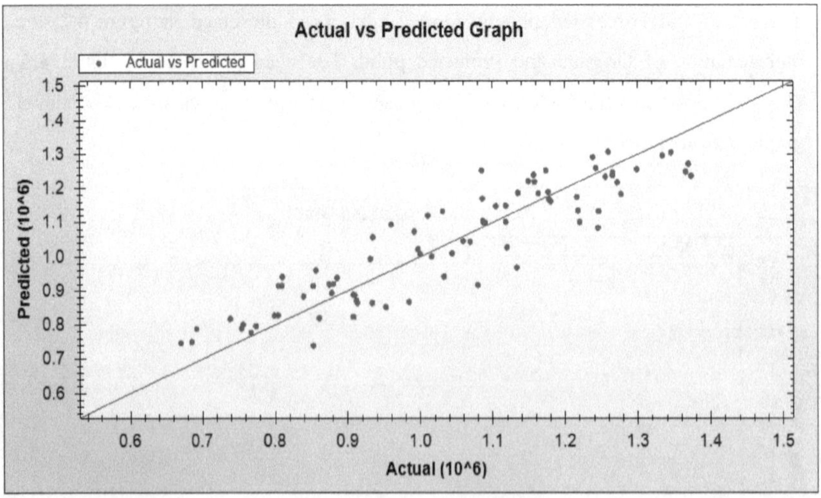

Figure 4.4 Representation of Original vs Projected for Air India air traffic

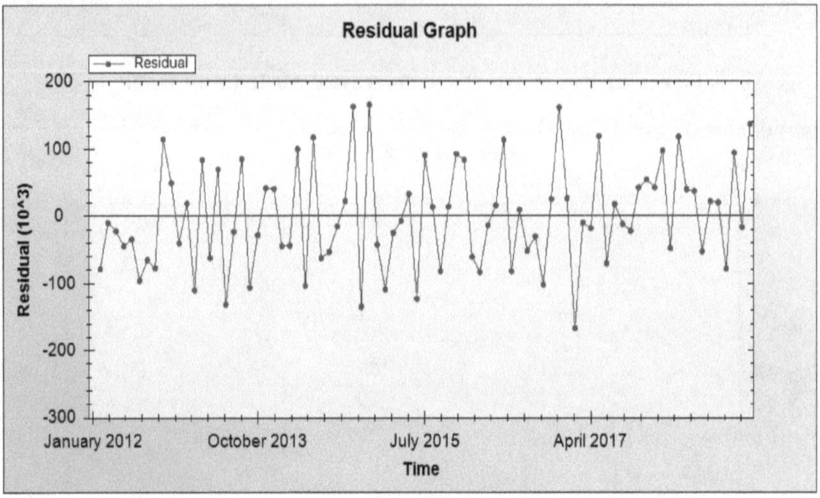

Figure 4.5 Plot of Residuals of Air India air traffic

4.6.2 Artificial Neural Networks Architecture for INDIGO

In the investigation, Multi-Layer Perceptron (MLP) network has been used for the prediction of air traffic flow. For the development of the Artificial Neural Network (ANN) model, 84 months of data points regarding the number of passengers travelled by Indigo domestic services have been taken. Air traffic was considered as an input variable, and one hidden layer was considered. Different Artificial Neural Network (ANN) models have been developed on the training data set. In the present study ERROR, MSE, RMSE, and MAE values have been used to evaluate the performance of the model and predicted results. The specification of all the models has been presented in table 4.2.

4.6.2.1 Numerical Results

It can be observed from table 4.2 that model 4, i.e., a neural network with 4 hidden neurons, has minimum Error, Mean Square Error (MSE), Root Mean Square Error (RMSE), Mean Absolute Error (MAE) values. Hence it is used to forecast the future air traffic.

Model	Hidden layer	Hidden neurons	ERROR	MSE	RMSE	MAE
M1	1	1	0.05943	43382904160.434	208285.631	162845.40161
M2	1	2	0.05905	43080049235.580	207557.34	162061.2570
M3	1	3	0.05670	41372540352.891	203402.41	157973.9488
M4	1	4	0.05332	38883022272.473	197187.784	152643.5503

Table 4.2 Different Neural Network models' ERROR, MSE, RMSE and MAE values of Indigo

Actual and Forecasted plot by model 4 has been presented in figure 4.6, the plot of Actual and Predicted values has been presented in figure 4.7, Actual versus Predicted plot has been presented in figure 4.8 as well as Residual graph presented in 4.9.

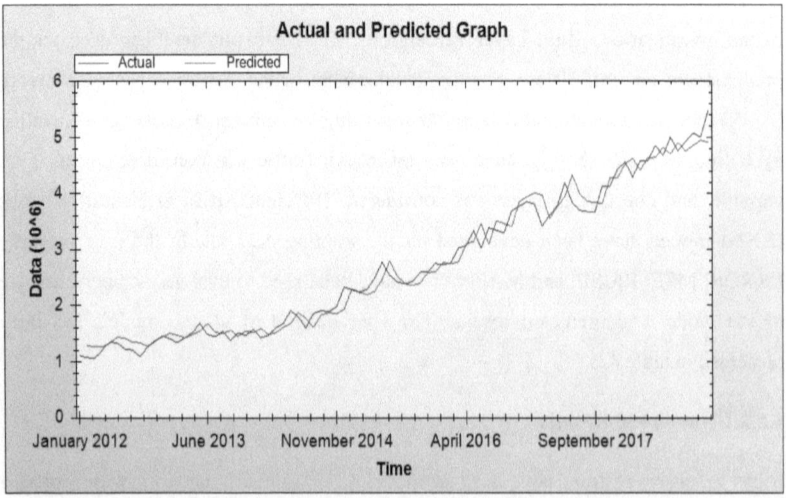

Figure 4.6 Representation of Original and Projected Indigo air traffic

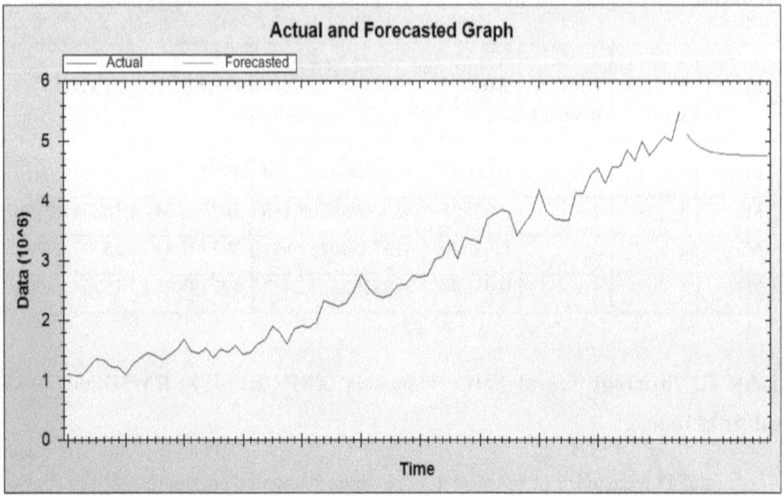

Figure 4.7 Representation of Original and Forecasted Indigo air traffic

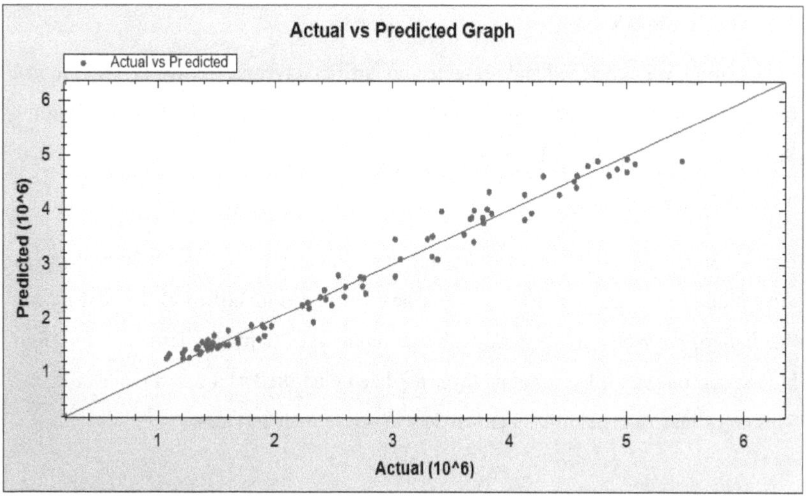

Figure 4.8 Representation of Original vs Projected for Indigo air traffic

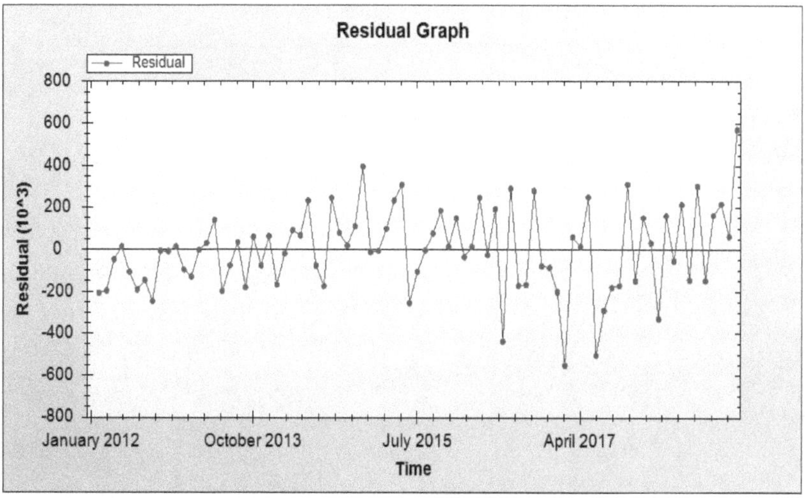

Figure 4.9 Plot of Residual of Indigo air traffic

4.7 CONCLUSION

In this study, we developed and presented an Artificial Neural Network (ANN) model for forecasting the air traffic of Air India and Indigo domestic Services by using the Multi-Layer Perceptron (MLP) network. The results exhibited by the model were quite satisfactory.

For Air India, the ANN architecture with one input layer, 1 hidden layer carrying three hidden neurons as well as 1 output layer has been identified as the best model. For Indigo, the ANN architecture with one input layer, 1 hidden layer carrying four hidden neurons as well as 1 output layer has been identified as the best model.
This investigation is helpful for air carriers to revise their services.

CHAPTER-V

TIME SERIES ANALYSIS FOR LONG MEMORY PROCESS OF AIR TRAFFIC USING ARFIMA

5.1 INTRODUCTION

Air transport is currently one of the fastest growing service sectors. It is of great importance to analyze and forecast air traffic in designing development strategies for air carriers. Presently the issue is to notice air traffic flow having characteristics of self-similarity to analyze and depict some executing model to air traffic. Previously the road traffic has been modelled using the Poisson approach (Leland et al., 1994), same has been practiced in Ethernet, LAN, WAN, and WWW traffic. In general, most of the times nature of the traffic is bursty. This traffic congestion may be studied mathematically as LRD (long range dependence) or self-similarity. The Poisson process cannot be applied to study the self-similar behaviour in traffic (Allan T Anderson et al., 1998). Using MAP (Markovian arrival process), which studies self-similarity behaviour in data, can be employed for required time scales through balancing second order sample constants from the series.

The theme of the immediate section is to scrutinize the nature of the Self -Similar Behaviour of Air Passengers Arrival Pattern of Air Service providers and then to fit an appropriate time series model to forecast air traffic. In the study, air passengers' data has been contemplated being traffic also established that the above mentioned traffic is assuring features of self-similarity. At this point, we computed the Hurst index by using various methods.

In chapter 3, we have discussed the ARIMA model to forecast air traffic. The approach of ARIMA analyzes the short memory theme. Long memory processes are analyzed by ARFIMA (Autoregressive Fractionally Integrated Moving Average) approach, which have stabilized mean and variance that is stationary. ACFs of the process with long memory deteriorate very steadily than the process with a short memory. This long Range Dependence (LRD) can be best analyzed by ARFIMA (or FARIMA) models, which also satisfies the principle of parsimony.

5.2 METHODS OF ESTIMATING THE HURST EXPONENT OF SELF SIMILAR SYSTEM

Measuring self-similarity mathematically is not easy to depict and interpret due to its extreme level of complexity. The Hurst (Hurst, 1951) parameter is enchanting since it narrates too many areas of mathematics example: autocorrelation, fractals, wavelets, etc. Hurst index proposes an extensive measure of whether a time series has LRD or not. This has been beneficial in the survey on network traffic analysis and modelling. The Hurst index H was employed to approximate the self-similarity, the Hurst exponent H (Hurst, 1951), was named later than the hydrologist, who depleted so many years to examine the difficulty of storage of the water, also conclude the altitude forms of the Nile River. The range of Hurst exponent is $0.5 \leq H < 1$. Measuring exponent H is a complicated task. We have many techniques for measuring Hurst exponent, the methods illustrate considerably different results; some of the methods are rescaled adjusted range Statistics, Periodogram method, Correlogram method, residuals of regression method, absolute moment's method, Higuchi's method. Despite identical speculative foundations, the practical application of these methods depends on the nature of the reviewed case. The calculation is expensive in terms of computational power and time.

We introduced a new method that depends on percentiles is expected to compute the Hurst parameter. A percentile in Statistics means the variable resolution under that a specific portion of data points dropdown. We have no typical description of percentile (Lane and David, 2007), as the number of observations is large, in that case, all definitions yield similar results.

5.2.1 RESCALED ADJUSTED RANGE Statistics

Self-similarity statistically signifies the statistical properties and is designed for the whole data sets. These are identical to the sub-sections of each data set, which is being relevant toward measuring the Hurst parameter, where the Rescaled range statistics are calculated (Gospodinov and Evgeniya Gaspodinova, 2005) over divisions of various sizes. Figure 5.1 describes that the Rescaled range is computed about the whole series ($RSave_0 = RS_0$). After that, it is computed for bisects of two data sets, ensuing RS_0 and RS_1. Procedure persists by separating each one of the preceding segments in half and manipulative RS of every one of the novel segment and averaged the RS resolution of every segment. The subsection ends as the section acquires very small (as a minimum of eight data points).

The Hurst index H is approximated by measuring the mean of rescaled range over several sections of data. Designed of known observations as a set, $X_1, X_{2,\ldots\ldots}, X_n$ by the average of the sample μ equals to $E[X]$ be described a series of regulated partial aggregates

$$W_j = (X_1 + X_2 + ::: + X_j) - jX(n), j = 1,2,3\ldots\ldots n \tag{5.1}$$

Where $\overline{X}$ is the average of the first n number of observations.

$R(n)$ is the range described below.

$$R(n) = \max(0, W_1, W_2, \ldots\ldots, W_n) - \min(0, W_1, W_2, \ldots\ldots, W_n) \tag{5.2}$$

$S(n)$ is the standard deviation on data points $X_1, X_2, \ldots, X_n$ is established as

$$S(n) = \sqrt{E(X_i - \mu)^2} \tag{5.3}$$

The Hurst index be accessible through adjusted rescaled range

$$\text{Statistic R/S} = \frac{R(n)}{s(n)} \tag{5.4}$$

A predictable realization from $R(n) = S(n)$ gratify the relation of power principle asymptotically

$$E\left[\frac{R(n)}{S(n)}\right] \rightarrow cn^H, \text{ as } n \rightarrow \infty, \tag{5.5}$$

Where a finite constant $c > 0$ and $H > 0.5$ is Hurst index.

Through a power function with an index of 0.5, the predictable value will be described for the short range-model.

$$E\left[\frac{R(n)}{S(n)}\right] \rightarrow dn^{0.5}, \text{ as } n \rightarrow \infty, \tag{5.6}$$

<table>
<tr><td colspan="16">RS_0</td><td>$RSave_0$</td></tr>
<tr><td colspan="8">RS_0</td><td colspan="8">RS_1</td><td>$RSave_1$</td></tr>
<tr><td colspan="4">RS_0</td><td colspan="4">RS_1</td><td colspan="4">RS_2</td><td colspan="4">RS_3</td><td>$RSave_2$</td></tr>
<tr><td colspan="2">RS_0</td><td colspan="2">RS1</td><td colspan="2">RS2</td><td colspan="2">RS3</td><td colspan="2">RS_4</td><td colspan="2">RS_5</td><td colspan="2">RS_6</td><td colspan="2">RS_7</td><td>$RSave_3$</td></tr>
<tr><td>RS_0</td><td>RS_1</td><td>RS_2</td><td>RS_3</td><td>RS_4</td><td>RS_5</td><td>RS_6</td><td>RS_7</td><td>RS_8</td><td>RS_9</td><td>RS_{10}</td><td>RS_{11}</td><td>RS_{12}</td><td>RS_{13}</td><td>RS_{14}</td><td>RS_{15}</td><td>$RSave_4$</td></tr>
</table>

Figure 5.1: Estimating the Hurst parameter

Here d is the bounded value. The variation among Equation (5.5) as well as Equation (5.6) is called the Hurst effect.

5.2.2 VARIANCE-TIME method

In accordance with variance based description given that

$$\mathrm{Var}[X^m] = \frac{Var\,[X]}{m^\beta} \tag{5.7}$$

Where $H=1-\frac{\beta}{2}$

When aggregate a process, through various stages of aggregation X_m, the variance is generally declining with H=0.5. The exemption, though, is in processes self-similar, wherever the variance is lessening gradually, in proportion toward a power principle (with high values of H).

The Equation (5.7) is able to modified as

$$\log Var[X^m] \sim \log[Var(X)] - \beta \log(m) \tag{5.8}$$

Here $log\ Var[X^m]$ is not depending on X_m and also constant, when plotting $Var[X]$ against m on a log-log scale, through deriving a straight line by least-squares by resulting points, the consequence is supposed to be line equation with slope $-\beta$ and the Slope estimate falls in (-1,0) offered self-similarity.

5.2.3 PERIODOGRAM Method

(Brockwell and Davis, 1996) proposed the Periodogram method. By way of its spectral depiction, studying the time series is simply the study of its stationary process in the frequency domain. The Periodogram is specified by (Brockwell and Davis, 1996)

$$I_N(\lambda) = \frac{1}{2\pi N}\left|\sum_{j=0}^{N-1} X_j e^{j\lambda}\right|^2 \quad (5.9)$$

From Equation (5.9), λ denotes the Fourier frequency, and N tells the count of observations of time series, as well as X_j indicates series for the dataset. In the direction of approximate H value first, we approach Periodogram. As $I_N(\lambda)$ is spectral density estimator, the LRD dataset is supposed to own a Periodogram and is in proportion with $|\lambda|^{1-2H}$ nearby zero. Subsequently, a throwback of the log of Periodogram and a log of frequency λ is required to provide a coefficient of $1-2H$.

5.2.4 CORRELOGRAM Method

The plot of Autocorrelation Function (ACF) in time series analysis is acknowledged like Correlogram, where the approximated correlation is able to be specified in terms of ACF of $\gamma(k)$ as (Brockwell and Davis, 1996)

$$\rho(k) = \frac{\gamma(k)}{\gamma(0)} \quad (5.10)$$

It is examined as a slow root of correlation, and proportional to k^{2H-2} for $0.5 < H < 1$ specify Long-memory process. As a result of the sample, ACF is supposed to show this property. An enhanced plot for performing LRD is the plot for the Auto Correlation Function within the logarithmic scale. For long memory processes, if a deterioration of correlation asymptotically is hyperbolic, subsequently the values in the graph supposed to roughly spread in the order of a line through a -ve slope of $2H$-2, the points must be inclined to diverse of minus infinity on an exponential tempo for short memory process. If the series has LRD, then the log-log Correlogram is very helpful. Since a preface heuristic approach of data, it is a realistic one. Some difficulty about sample correlation exists, which is not as much recognized can be establishing

(Beran et al., 1995; Beran, 1994). Although it is neither an extensively used nor striking process of assessment, still the self-similarity measuring index H can be approximated with this method. Obtaining the form of an equation is

$$\rho(k) = \hat{H}(2\hat{H}-1)k^{2\hat{H}-2} \tag{5.11}$$

5.2.5 RESIDUALS OF REGRESSION Method

(Montgomery et al., 2001), in the Residual of Regression, the First step is to segregate the series taken with a number of blocks, say size m., and for each block m, the partial sum of series are designed for Y_i, i=1,2,3,…,m. Obtain a straight line in the direction of Y_i also calculate the error variation. And the process is recurring in every segment also, resultant variances of a dataset are centred. Thus, when the outcome is graphed on a log-log scale against *m* and is supposed to acquire a line with slope 2*H*, i.e. the variance of residuals against the number of blocks.

5.2.6 HIGUCHI'S Method

A technique recommended by (Higuchi 1988) is extremely comparable to the absolute moments' method by $n = 1$. It makes use of

$$L(m) = \frac{N-1}{m^3} \sum_{i=1}^{m} \left[\frac{N-1}{m}\right]^{-1} \sum_{k=1}^{[(N-i)/m]} \left| \sum_{j=i+(k-1)m+1}^{i+km} X(j) \right| \tag{5.12}$$

Now, *m* is the equivalent of block size, *N* is the time series length, which indicates the greatest integer function. For Long-Range Dependence (LRD) series $E(L(m)) \sim cm^{H-2}$. Basically, the variation among the absolute moments method stretches out using a sliding window to calculate aggregate series in-stead of using non intersecting blocks. Since the method is a lot more computationally rigorous. Though this adjustment may consequences in improved precision in the sorter time series, there seems to be no gain in using it for the series length of 10000. More details of this method are discussed in (Taqqu and Teverovasky, 1997; Taqqu et al., 1996)

5.2.7 PERCENTILE Method

We propose this method as follows:

Percentile is a variable value under which a definite number of observations fall. We have no typical description of percentile (Lane and David, 2007), though all descriptions relent similar outcomes, while the observations are incredibly large. One classification of percentile is repeatedly specified in books, is that the Pth percentile ($0 \leq P \leq 100$) of N number of values are getting hold through first operating the rank.

$$n = \frac{P*N}{100} + \frac{1}{2} \tag{5.13}$$

For given data (t, X_t), $t \geq 0$. Initially, it is needed to get percentiles

(P_i, $i = 1,2,3,\ldots,100$) designed for a specified time series data using the formula

$$P_i = \frac{i*N}{100} + \frac{1}{2}; \quad i = 1,2,\ldots,100 \tag{5.14}$$

$P_i = i^{th}$ percentile is an individual way of mean values in numerical data as quartiles (Q_1, Q_2, Q_3). Then plot log-log plot of percentile, i.e. percentile number vs percentiles. A straight line equation $X_t = \beta t + c$ is attained. Then compute Hurst exponent H by with the equation $H = 1\text{-}\beta/2$

The advantage of this method is exploited for a large set of infinite traces and provide theoretical justification for the aggregation window(buffer) size using the self-similar model (Webhostingtalk forum, 2008). This method is applied to road traffic of a national highway in India and is validated with other methods, namely Periodogram method, Correlogram and residual of regression method in the manuscript (Malla Reddy Perati et al., 2012)

5.3 AUTOREGRESSIVE FRACTIONALLY INTEGRATED MOVING AVERAGE (ARFIMA) PROCESS

ARFIMA or FARIMA models are extensively employed in diverse disciplines like economics and hydrology. Long-Range Dependence (LRD) has been primarily introduced by Hurst (1951) in the field of hydrology. Hosking (1981) represented Long-Range Dependence in hydrology by ARFIMA. Granger and Joyeux (1980) used ARFIMA to represent Long-Range Dependence in economics. LRD can be conveniently measured by the Hurst parameter 'H'. Various methods of calculating Hurst parameter 'H' and the relationship between self-similar and LRD are discussed above.

The ARFIMA models are the generalization of ARIMA models. Where'd' is allowed to take non integer values, $-0.5 < d < 0.5$. This fractional integrating parameter'd' is used to capture the reactions for a long period. In comparison, the parameters of ARMA seize the reactions in a short period.

According to Baillie (1996), Hosking (1981), and others, the FARIMA process having $d < 0$ indicates intermediate memory process and $d > 0$ indicates long memory process. According to Box-Jenkins, Reinsel (2008, 429), the ARFIMA process with $d < 0$ indicates a long memory process as they decay slowly at a hyperbolic rate. Thus we follow in this study the ARFIMA process with $-0.5 < d < 0.5$ indicates a long memory process.

The ARFIMA model is

$$\emptyset(B)(1-\mathrm{B})^{d}\, Z_t = \theta(B)a_t \tag{5.15}$$

Where

$\emptyset(B) = 1 - \emptyset_1 B - \emptyset_2 B^2 - .\dots - \emptyset_p B^p$ is AR operator, $\theta(B) = 1 - \theta_1 B - \theta_2 B^2 - .\dots - \theta_q B^q$ is MA operator and d is the fractional integrating parameter, $-0.5 < d < 0.5$.

$$(1-B)^d = \sum_{i=0}^{\infty}(-1)^i \frac{\Gamma(i+d)}{\Gamma(i+1)\Gamma(\mathrm{d})} B^i \tag{5.16}$$

Γ is the gamma function.

5.4 MATERIALS AND METHODS

Air passengers' data refer to the number of passengers travelling monthly by Air India and Indigo on scheduled Domestic services for the years 2012 to 2018, which explored in the introduction. The data has been taken from the official DGCA website. Air passengers monthly travel data by Air India and Indigo domestic flights have been considered. In this research, air passengers' dataset has been contemplated being traffic also established that the above mentioned traffic is assuring features of self-similarity. At this point, we computed the Hurst index by using various methods, and then the ARFIMA or FARIMA model has been fitted to air traffic.

5.5 ANALYSIS OF LONG RANGE DEPENDENCE OF AIR INDIA AIR TRAFFIC

5.5.1 Investigating Self-similarity of Air India air traffic

The dataset has to be deliberated towards shallow frequencies or extreme lags where hardly any measurements are accessible. The index H possesses a range of $0.5 < H < 1$. Computing index H is an assignment. Diverse approaches are accessible to determine the intensity of self-similarity in a dataset and forecasting analysis. Three procedures are placed to enumerate the Hurst index: Variance-time analysis, Correlogram, and one more approach built on percentiles are appealed after that collated with the mentioned approaches.

5.5.2 ARFIMA or FARIMA Model

By Self Similarity methods, it has been found that Air India air traffic data has exhibited LRD (Long Range Dependence). The LRD data has stabilized mean and variance, i.e., stationary, and it's ACFs degrade very gradually compare to Short Range Dependent (SRD) data. The ARFIMA (FARIMA) approach gives an optimal model for a process with LRD parsimoniously. The ARFIMA approach is an addendum of the ARIMA approach that allows a fractional difference $-0.5 < d < 0.5$ to model long period effects. The time series with LRD can be best analyzed by the ARFIMA procedure. So FARIMA technique has fitted to the Air India air traffic data. The STATA software has been used for this analysis. ARFIMA parameters have been estimated using the Modified Profile Likelihood (MPL) method. The fractional difference parameter estimated by MPL will have a minimum bias of a small sample compared to the MLE.

The ARFIMA model parameter estimates have been presented in table 5.1.

		Parameter	Estimate
ARFIMA(p, d, q)	AR1	Ø	0.657
	MA1	Θ	-0.999
	Fractional difference	D	-0.347

Table 5.1 ARFIMA Model Parameters of Air India

5.5.3 Numerical Results

Hurst exponent H gives the measurement of self-similarity. We provided three strategies to determine the Hurst exponent Correlogram method, Variance-time analysis also an additional method on the base of percentiles is well-designed and measure through the supposed methods. Using the Correlogram method, the value of Hurst index H is worked out for data mentioned in section 5.5.1 the acquired H value is 0.839. From the variance-time analysis method, the attained value of Hurst H is 0.792, and by using the Percentiles method, the estimated H is 0.797.

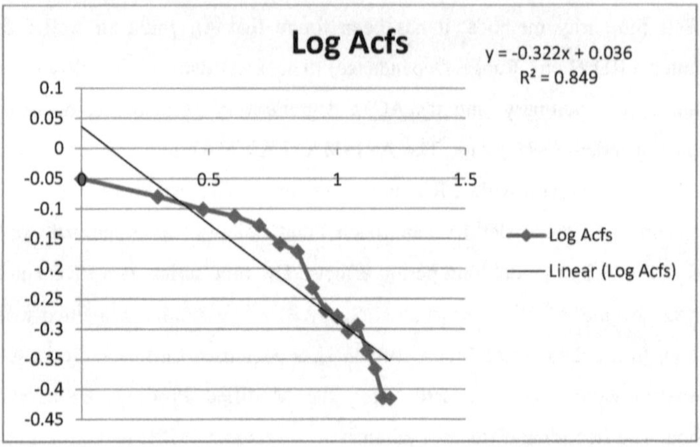

Figure 5.2 ACF verses Lag of Air India air traffic.

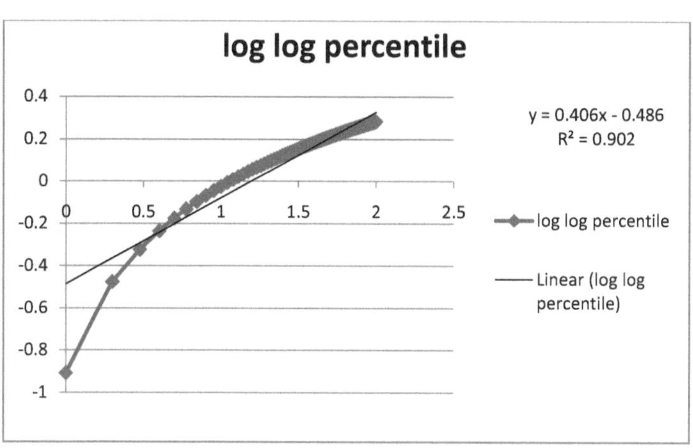

Figure 5.3 Percentile versus percentile number of Air India air traffic

As the data exhibited self-similar behaviour, the ARFIMA model has been fitted to the Air India air traffic data. The model parameters are as follows. Ø, the AR component has been evaluated as 0.657; θ, the MA component has been evaluated as -0.999, and the fractional difference parameter d was estimated to be -0.347.

Performance evaluation measures are presented in table 5.2

Measure	Value
R-squared	0.948
RMSE	43588.99
MAE	36854.83
MAPE	3.248

Table 5.2 Performance evaluation measures of ARFIMA (1,-0.347,1) of Air India

The representation of original and forecasted points of ARFIMA (1,-0.347,1) has been presented in 5.4.

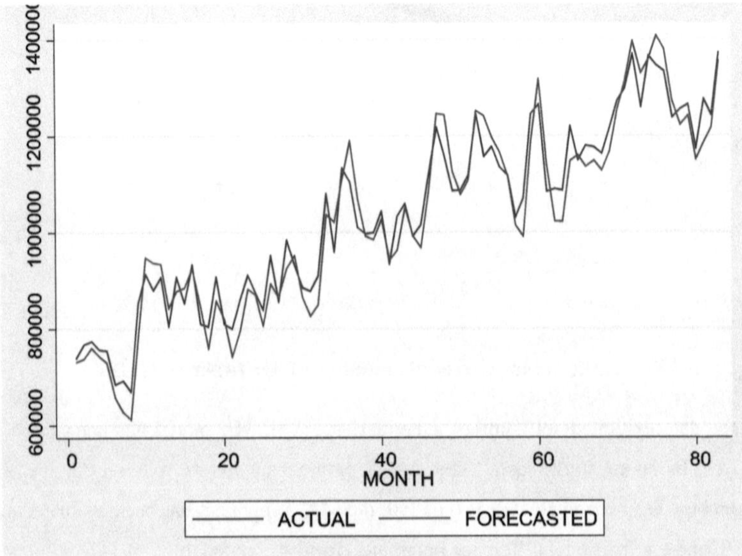

Figure 5.4 Representation of Original and Forecasted points of ARFIMA of Air India

5.6 ANALYSIS OF LONG RANGE DEPENDENCE OF INDIGO AIR TRAFFIC

5.6.1 Investigating Self-similarity of Indigo air traffic

The theme of this section is to examine the nature of the Self -Similar behaviour of Air Passengers Arrival Pattern of Indigo Services. In this research, air passengers' dataset has been contemplated being traffic also established that the above mentioned traffic is assuring features of self-similarity. At this point, we computed the Hurst index by using Correlogram, Variance Time Plot and Percentile methods.

5.6.2 ARFIMA Model

In 5.6.1 by Self Similarity methods, it has been found that Indigo air traffic data has exhibited LRD. The LRD data has stabilized mean and variance, i.e., stationary, and it's ACFs degrade very gradually compare to Short Range Dependent (SRD) data. The ARFIMA (FARIMA) approach gives an optimal model for a process with LRD parsimoniously. The ARFIMA approach is an addendum of the ARIMA approach that allows a fractional difference $-0.5 < d < 0.5$ to model long period effects. The time series with LRD can be best analyzed by the ARFIMA procedure. So FARIMA technique has fitted to the Indigo air traffic data. The STATA software has been used for this analysis. ARFIMA parameters have been estimated using the Modified Profile Likelihood (MPL) method. The fractional difference parameter estimated by MPL will have the minimum bias of a small sample compared to the MLE.

The ARFIMA model parameter estimates have been presented in table 5.3.

		Parameter	Estimate
ARFIMA(p, d, q)	AR2	Ø	0.637
	MA2	Θ	-0.715
	Fractional difference	D	-0.478

Table 5.3 ARFIMA Model Parameters of Indigo

5.6.3 Numerical Results

Hurst exponent H gives the measurement of self-similarity. We provided three techniques to compute the Hurst exponent Correlogram method, Variance-time analysis also an additional method on the base of percentiles is well-designed and measure through the supposed methods. Using the Correlogram method, the value of Hurst index H is worked out for data mentioned in section 5.6.1 the acquired H value is 0.859. From the variance-time analysis method, the attained value of Hurst H is 0.802, and by using the Percentiles method, the estimated H is 0.79.

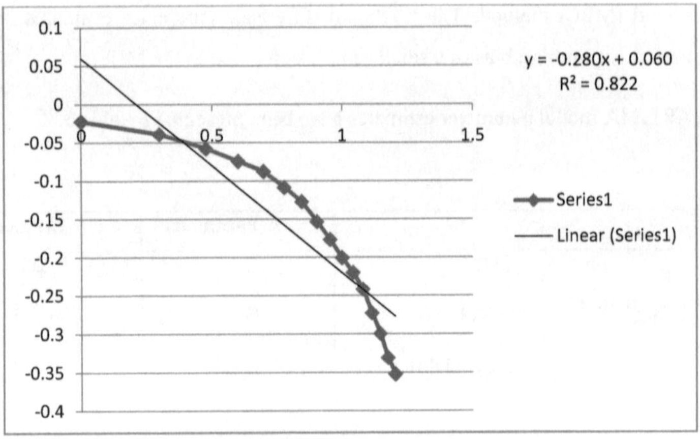

Figure 5.5 ACF verses Lag of Indigo air traffic.

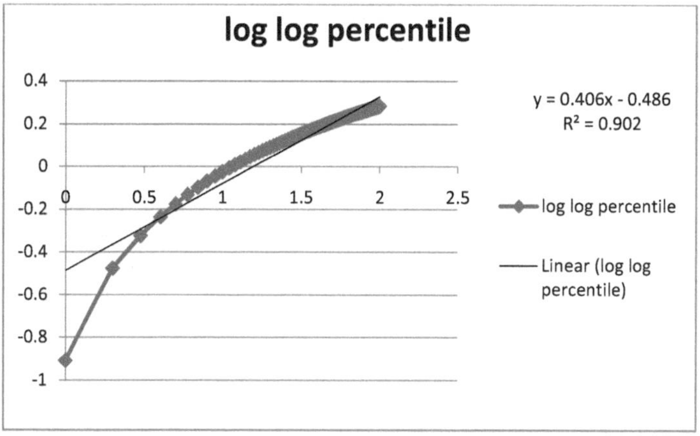

Figure 5.6 Percentile versus percentile number of Indigo air traffic

The AR component Ø has been determined as 0.637, the component of MA, θ, has been found as -0.715, and the fractional difference parameter d was estimated to be -0.478.

Performance evaluation measures are presented in table 5.4

Measure	Value
R-squared	0.995
RMSE	98691.44
MAE	78165.45
MAPE	2.01

Table 5.4 Performance evaluation measures of ARFIMA (1,-0.347,1) of Indigo

The representation of original and Forecasted points of ARFIMA (2,-0.478,2) has been presented in 5.7.

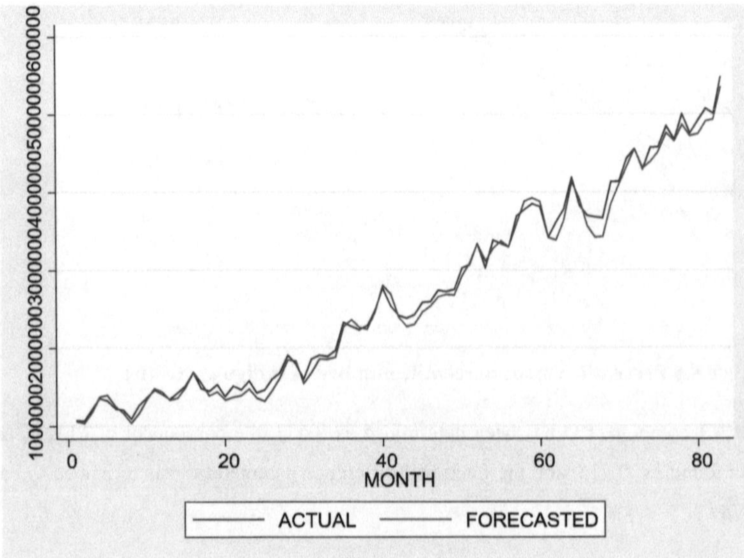

Figure 5.7 Representation of Original and Forecasted points of ARFIMA of Indigo

5.7 OVERALL RESULTS

5.7.1 Comparison of Adequacy of the Models fitted to Air India air traffic data

Optimum model out of entire fitted approaches of Air India air traffic data has been identified using R Square, MAE, RMSE, and MAPE values presented in table 5.5.

Method	RMSE	MAE	MAPE	R Square
ARIMA	65389.581	51419.953	5.084	0.882
SARIMA	78229.883	61700.147	5.980	0.767
ANN	76728.6215	63805.4	6.315	0.826
HOLT-WINTERS' ADDITIVE	51298.805	41178.488	4.116	0.926
ARFIMA	**43588.99**	**36854.83**	**3.248**	**0.948**

Table 5.5 Performance evaluation measures of all the models of Air India air traffic

From table 5.5, ARFIMA (or) FARIMA model gives the best fit to the Air India Scheduled domestic air traffic data as RMSE, MAE and MAPE values of ARFIMA are less than other models, and also R Square of ARFIMA is greater than other models.

5.7.2 Comparison of Adequacy of the Models fitted to Indigo air traffic data

Optimum model out of entire fitted approaches of Indigo air traffic data has been identified using R Square, MAE, RMSE, and MAPE values presented in table 5.6.

Method	RMSE	MAE	MAPE	R Square
ARIMA	181805.943	143238.405	5.680	0.980
SARIMA	149450.052	116813.034	4.228	0.984
ANN	197187.784	152643.5503	6.151	0.975
HOLT-WINTERS' ADDITIVE	118966.697	92660.607	4.010	0.991
ARFIMA	**98691.44**	**78165.45**	**2.01**	**0.995**

Table 5.6 Performance evaluation measures of all the models of Indigo air traffic

From table 5.6, ARFIMA (or) FARIMA model gives the best fit to the Indigo Scheduled domestic air traffic data as RMSE, MAE and MAPE values of ARFIMA are less than other models, and also R Square of ARFIMA is greater than other models.

5.8 CONCLUSION

In this chapter, the number of air passengers considered as air traffic. Air traffic of Air India and Indigo has been considered for this investigation. For these two air traffic, we have verified self-similar behaviour by determining the Hurst index by various methods. The obtained H values are almost similar to each other. By these values, the air traffic of Air India and Indigo has proven to have LRD (Long Range Dependence). Thus the ARFIMA model was fitted to both the air traffic. The performance evaluation measure and the plot of actual and forecasted points of the approach were also presented in the chapter. Also, overall comparisons of ARFIMA with ARIMA, SARIMA, ANN and Holt-Winters additive models are placed. This comparison showed ARFIMA or FARIMA as the best fit for these two air traffic datasets. This analysis is useful for air service providers to design and modulate their Quality of Services (QoS).

CHAPTER VI

SUMMARY AND CONCLUSIONS

In this chapter, an overall summary, conclusions and possible open problems are placed.

6.1 SUMMERY

In the modern world, many researchers are working on various real-time problems which are connected to time series forecasting models, however, despite the importance, the Self-similar nature not been addressed. Many approaches are developed for forecasting time series. This thesis concentrated on analyzing the Long Memory Process, this can be done by taking account of Self-similar nature.

The aspiration of this analysis is to explore the forecasting model for the air traffic of Air India and Indigo Scheduled Domestic flights. Innumerable approaches of time series like ARIMA, SARIMA were employed to the air traffic data. Exponential smoothing methods, which have broad applications in time series forecasting, also applied to the air traffic data. Another alternative approach Artificial Neural Networks employed to develop a paradigm for forecasting air traffic.

The thesis also aims to investigate the self-similar structures in air traffic. Also, the concept of self-similarity has now proved itself to be extraordinarily useful. This research explains the current state of knowledge regarding the self-similar process and their applications; concept, definitions and basic properties are emphasized. The fractal property Self-similarity gives a direction to the researchers as a road map of the area of self-similarity that permits for further investigation.

This thesis is an assortment of chapter contributions that fetches mutually appropriate past works on both sides of a section of themes covering traffic extent, modelling, traffic control, and performance analysis for self-similar data.

The main reason of this thesis is to present an understanding of self-similarity nature and yet solid description of the important progress and effects relating to self-similar traffic and its different aspects, with an endeavour of portion as a contemplative object that confines the modern in the ground. The thesis is structured approximately three primary subtopics, which are mainly air traffic forecasting using time series models, air traffic forecasting using artificial neural networks and analysis of LRD of air traffic using self-similarity as well as ARFIMA.

The present thesis intended on scheming the self-similarity using various methods by Hurst parameter. Air India and Indigo's air traffic prove themselves as a self-similar paradigm by means of disparate Hurst Index methods. Hence ARFIMA technique was operated on air traffic dataset also compared with the above said models ARIMA, SARIMA, Holt-Winters additive and ANN. ARFIMA model showed a better performance result as expected. Quantitative results obviously disclose that the analysis shown in this research supports the enhancement of traffic depiction to assure the quality of service guarantee.

6.2 CONCLUSIONS

Various forecasting models to air traffic data corresponding to Air India and Indigo Scheduled Domestic services have been studied. The forecasted models discussed were stochastic models, exponential smoothing models, ANN models. The self-similarity of air traffic has investigated, and then the ARFIMA or FARIMA model was employed.

Broadly chapter I is the fundamentals of the domain of research. We discussed the role of Indian civil aviation along with the past and progress. Industrial profiles of Air India and Indigo have been placed. A literature survey is made in this chapter.

In chapter II, We discussed basic concepts of time series analysis, preliminaries of self-similarity, with introduced the concept of self-similarity from various aspects.

In chapter III, various stochastic models and exponential smoothing models for forecasting the time series have been discussed. Appropriate stochastic model and Holt-Winters additive model have been fitted to Air India, and Indigo scheduled domestic air traffic. The best model has been identified using performance evaluation metrics. The additive approach of Holt-Winters has been adopted as a foremost approach for these two air traffic data. Numerical results, the plots exhibiting the relation of actual, predicted and forecasted are presented in the chapter.

In chapter IV, the recent popular approach ANN (Artificial Neural Network) technique for forecasting the time series has been discussed. Various ANN models using Multi-Layer Perceptron (MLP) have been developed for Air India and Indigo Scheduled Domestic air traffic. For Air India, the ANN model with three hidden neurons was determined as the finest architecture using model fit statistics. For Indigo, the ANN model with four hidden neurons was picked as the finest architecture using model fit statistics. The plots of actual and predicted, actual versus predicted, actual and forecasted, and residuals are presented in the chapter.

In Chapter V, we discussed measures for finding Hurst exponent of self-similar progression. Here we proposed a model depend on Percentiles and conventional by existing methods such as Rescale Adjusted Range Statistics is projected. In this method the Hurst index is approximated by measuring the mean of rescaled range data. Also, the Variance-Time method and that of its aggregating a process for self-similarity are briefed. Estimating the spectral density that is a series with LRD showed as Peridogram method. In time series analysis, classification of the predictable correlation be able to specified in terms of ACF is generalized Correlogram method is used to fit the self-similar process. Higuchi's method proposed is related to the absolute moments used to show self-similarity and apply the residual of the regression method. A study on the self-similar behaviour of Air India and Indigo air traffic is discussed using the Hurst index by three methods Correlogram, Variance Time method and Percentile method. The H value by these methods indicated that the air traffic possesses LRD (Long Range Dependence).

As the air traffic data has LRD, ARFIMA (FARIMA) procedure is presumed to be the optimum procedure for the air traffic dataset. Thus ARFIMA (FARIMA) model has been fitted to Air India and Indigo air traffic. Overall comparisons have been presented at the end of the chapter. ARFIMA is compared with ARIMA, SARIMA, ANN and Holt-Winters additive models. Practically all the models are good for these two air traffic data sets as the MAPE of all the models are less than 10%. But finally, model fit statistics make known that ARFIMA (FARIMA) model is the best fit for air traffic data. Thus, when the time series data exhibits Long-Range Dependence (LRD), ARFIMA (FARIMA) gives the best performance results.

REFERENCES

A) Journals/Articles

1. Aalan,G., Konheim and Matin Reiser,A Queuing Model with Finite Waiting Room and Blocking, ACM Newyork,NY,USA,**Vol. 23, No. 2, pp. 328-341, (1976).**

2. Alberto Andreoni, Maria Nadia Postorino*, "Time Series Models To Forecast Air Transport Demand:A Study About A Regional Airport", 11th IFAC Symposium on Control in Transportation Systems Delft, The Netherlands, **August 29-30-31, 2006**.

3. Alberto Montanari and Renzo Rosso Murad S. Taqqu, "Fractionally dilferenced ARIMA models applied to hydrologic time series: Identification, estimation, and simulation" WATER RESOURCES RESEARCH, **VOL. 33, NO. 5, PAGES 1035-1044, MAY 1997.**

4. Allan, T. Anderson, Bo Friis Nielsen , A Markovian Approach for Modeling Packet Traffic with Long Range Dependence, IEEE Journal on Selected Areas in communications, **Vol. 16, No. 5, pp. 719-732, (1998).**

5. Adeleke, R.A., Ogunwale, O.D. and Halid, O.Y., Application of Queuing Theory to Waiting Time of Out-Patients in Hospitals, Pacific Journal of Science and Technology, **Vol. 10, No. 2, pp. 270-274, (2009).**

6. Andersen, A.T., Nielsen, B.F., A Markovian approach for modeling packet traffic with long range dependence, IEEE J.Select.Areas in Commun.,**Vol. 16, No. 5, pp. 719-732, (1998).**

7. Ashour M, Le-Ngoc T. Priority queuing of long-range dependent traffic, Computer communications, **Vol. 31, N0. 17, pp. 3954–3963, (2008).**

8. Baccelli, F., Boyer, p., and Hebuterne, G., Single-server queues with impatient customers, Advances in Applied Probability, **Vol. 16, No. 4, pp. 887-905, (1984).**

9. Beran, J., Taqqu, M.S. and Willinger, W., Long-Range Dependence in Variable Bit Rate Traffic. IEEE Transac-tions on Communications, **Vol. 43, No. 4, pp. 1566-1579, (1995a).**

10. Blinova T. O, Analysis of possibility of using neural network to forecast passenger traffic flows in Russia, ISSN:1648-7788 AVIATION, **Vol XI, No 1, 2007**

11. Chinwen cheong, Self-similarity in financial markets: A fractionally integrated approach, Mathematical and Computer Modelling,**Vol. 52, No. 4, pp. 459-471, (2010).**

12. Claudimar Pereira Da Veiga, Cássia Rita Pereira Da Veiga, Anderson Catapan, Ubiratã Tortato, Wesley Vieira Da Silva, "Demand forecasting in food retail: a comparison between the Holt-Winters and ARIMA models", Wseas Transactions On Business And Economics, E-ISSN: 2224-2899, **Volume 11, 2014**.

13. Chitra,N , Ra.Shanmathi , R.Rajesh, "Application Of Arima Model Using Spss Software - A Case Study In Supply Chain Management" International Journal of Science, Technology & Management, **Volume No 04, Special Issue No. 01, April 2015.**

14. Crovella, M. and Bestavros, A., Self-Similarity in World Wide Web Traffic: Evidence and Possible Causes., IEEE/ACM Transactions on Networking, **Vol. 5, pp. 835-846, (1997).**

15. Dheeraj Duhan , Nishant Arya, Prateek Dhanda, Lalit Upadhyay, k., Mathiyazhagan,Application of Queuing Theory to Address Traffic Problems at a Highway Toll Plaza, Applied Mechanics and Materials, **Vols. 592-594, pp. 2583-2587, (2014).**

16. Doodipala, M.R, Second Order Statistics of Time Series of Various Real Time Problems in Conjunction with Periodogram Technique, International Journal of Latest Trends in Engineering and Technology, **Vol. 3, No. 1, pp. 334-345, (2013).**

17. Erlang, A.K., The theory of probabilities and telephone conversations, Tidsskrift Matematika, **Vol. 20, pp. 33-39, (1909).**

18. Erramalli, A., Narayan,O. and Willinger, W., Experimental queuing analysis with long range dependent packet traffic, IEEE/ACM Trians Network, **Vol. 4, No. 2, pp. 209- 223, (1996).**

19. Etebong P. Clement, "Using Normalized Bayesian Information Criterion (Bic) to Improve Box - Jenkins Model Building" American Journal of Mathematics and Statistics **2014, 4(5): 214-221 DOI: 10.5923/j.ajms.20140405.02.**

20. Faraway J., Chatfield .C., "Time series forecasting with neural networks: a comparative study using the airline data", **Applied Statistics 47 (1998), pages: 231–250.**

21. Field A.J., Uli Harder & Peter Harrison, Measurement and modelling of self-similar traffic in computer networks, IEE Proceedings – Communications, **Vol. 151 , No.4 , pp 355-363,(2004).**

22. Fei Xue, Jiakun Liu, Yantai Shu, Lianfang Zhang, "Traffic Modeling Based on FARIMA Models" **Engineering Solutions for the Next Millennium, 1999 IEEE Canadian Conference on Electrical and Computer Engineering (Cat. No.99TH8411).**

23. Galbraith J.W, Zinde-Walsh .V, "Autoregression-based estimators for ARFIMA models", CIRANO Working **Papers, No: 2011s-11, Feb 2001.**

24. Gooijer, J. G., Hyndman, R. J., 25 years of time series forecasting. International Journal of Forecasting, **Vol. 22, No. 3, 2006, pp. 443-473.**

25. Heidemann D., Queue Length and Delay Distributions at Traffic Signals, Tranportation Research Part B Methodological, **Vol. 28,No. 5, pp. 377-389, (1994).**

26. Higuchi, T., Approach to an irregular time series on the basis of the fractal theory, Elsevier Science Publishers, **Vol. 31 No. 2, pp. 277 – 283,(1988).**

27. Hipel K.W, McLeod A.I, "Time Series Modelling of Water Resources and Environmental Systems", **Amsterdam, Elsevier 1994.**

28. Holt,C.C., Author'sretrospective on "Forecasting seasonal sand trends by exponentially weight moving averages"(Discussion),International Journal of Forecasting, **Vol. 20, No.11, 2004, pp. 13.**

29. Hyndman R. J., Koehler, A. B.,Ord, J. K., Snyder, R.D.,"Forecasting with exponential smoothing: the state space approach", Germany: **Springer. 2008, 351pp.**

30. Hornik K, Stinchcombe M, White H, "Multilayer feed-forward networks are universal approximators", Neural Networks 2 (1989), **pages: 359–366.**

31. Hurst H., Long-Term Storage of Reservoirs: An Experimental Study, Trans of the American Society of Civil Engineers, **pp.770-799, (1951).**

32. John H. Cochrane, "Time Series for Macroeconomics and Finance", Graduate School of Business, **University of Chicago, spring 1997.**

33. Kelly, F. P., Networks of Queues with Customers of Different Types. Journal of Applied Probability. **Vol. 12, No. 3 pp. 542–554, (1975).**

34. Kendall, D.G., Stochastic Processes Occurring in the Theory of Queues and their Analysis by the Method of the Imbedded Markov Chain, The Annals of Mathematical Statistics, **Vol. 24, No. 3, Pp, 323- 338, (1953).**

35. Kihoro J.M, Otieno R.O., Wafula .C, "Seasonal Time Series Forecasting: A Comparative Study of ARIMA and ANN Models", African Journal of Science and Technology (AJST) Science and Engineering Series **Vol. 5, No. 2, pages: 41-49.**

36. Kroner H., Comparative performance study of space priority mechanisms for ATM networks, Computer and Communication societies, **Vol. 3, pp.1136-1143, (1990).**

37. Leland W.E., Taqqu M.S., W. Willinger W. and Wilson D.V., On the Self-Similar Nature of Ethernet Traffic(Extended version) , IEEE / ACM Trans. Networking,**Vol. 2, No. 1, pp. 1-15, (1994).**

38. Lee .J, "Univariate time series modeling and forecasting (Box-Jenkins Method)", Econ 413, lecture 4.

39. Liu Chang, Qin Xiaowei, Zhang Sihai, Zhou Wuyang, Queuing Analysis of Self-Similar Traffic in Rayleigh Fading Channel, Networks Security Wireless Communications and Trusted Computing, **Vol. 1, pp. 287-291, (2009).**

40. Loynes R. M. , The stability of a queue with non-independent in inter-arrival and service times, Proc. Cambridge Philos. Soc, **Vol. 58, pp. 497-520, (1962).**

41. Lucantoni D.M., New results on the single server queue with a batch Markovian arrival process,Commun.Statist.-Statist.-Stochastic Models, **Vol.7, No. 1, pp. 1-46, (1991).**

42. Mala, Varma S.P., minimization of traffic congestion by using queuing theory, Journal of computer engineering,**Vol. 12, No. 1, PP 116-122,(2016).**

43. Malla Reddy Perati, Raj Kumar, L.P., Mallikarjun Reddy, D.,Approximate Markovian Model for Delay Behavior of Internet Router under Self – Similar Traffic Input, Journal of Interdisciplinary Mathematics, **Vol. 14, No. 3, pp. 321-329, (2011).**

44. Malla Reddy Perati, Raghavendra K., Reddy H.K., Mallikarjuna Reddy D., Self Similar Behavior of High way Road Traffic and Occupancy Analysis at Toll Plazas, Journal of Transportation Engineering ASCE, **Vol. 138, pp. 1233-1238, (2016).**

45. Mallikarjuna Reddy,D., Rajaiah, D., Malla Reddy,P., Krishna Reddy,M., Delay Behavior of Internet Router under Self-Similar Traffic via Rational Approximations, Journal of Communication Networks and Distributed Systems, **Vol. 14, No. 2, pp.134-144, (2015).**

46. Mallikarjuna Reddy, D., Second Order Statistics of Time Series of Various Real Time Problems in Conjunction with Periodogram Technique, International Journal of Latest Trends in Engineering and Technology (IJLTET), Vol. 3, No. 1, (2013).

47. Mandelbrot B.B., New Methods in Statistical Economics , Journal of Political Economy,**Vol.71, No.5, (1963).**

48. Mandelbrot, B.B and Ness, J.W. van., Fractional Brownian motion, Fractional noises applications, **Vol. 10, No. 4, pp. 422-437, (1968).**

49. Manish, K. Govil Michael C.Fu Queuing theory in manufacturing, Journal of Manufacturing systems,Vol. 18, No. 3 **pp. 214-240, (1999).**

50. Mannersalo, P, Norros I. A most probable path approach to queuing systems with general Gaussian input. Computer Networks, **Vol. 40, No. 3, pp. 399-412, (2002).**

51. Mohie El-Din M. M., Farag M. S.and Abouzeid A. A., Airline Passenger Forecasting in EGYPT (Domestic and International), International Journal of Computer Applications (0975 - 8887), **Volume 165 - No.6, May 2017**.

52. Martin Anokye, A.R. Abdul-Aziz, Kwame Annin, Francis T. Oduro, Application of Queuing Theory to Vehicular Traffic at Signalized Intersection in Kumasi-Ashanti Region, Ghana, American International Journal of Contemporary Research,**Vol.3,No.7,pp 23-28, (2013).**

53. Meng, Q. and Khoo, H.L.,Self-Similar Characteristics of Vehicle Arrival Pattern on Highways, Journal of Transportation Engineering, **Vol. 135, pp. 864-872, (2009).**

54. Nagatani, T., Self-similar behavior of a single vehicle through periodic traffic lights, Physica A, **Vol. 347, pp. 673–682, (2005).**

55. Naor P.and Yechiali U., Queuing problems with heterogenous arrivals and science, Operations Research. **Vol. 19, No. 3, pp. 722-734, (1971).**

56. Neuts M.F., A queue subject to extranuous phase changes, Advanced Application to Probability, **Vol. 3, No. 1, pp. 78-119, (1971).**

57. Neuts M.F., Versatile Markovian point process, Journal on Application to Probability, **Vol. 16, pp.764-769, (1979).**

58. Nicovandaele,Tom Van Woensel,Aviel Verbruggen, A Queuing Based Traffic Flow Model, Transportation Reasearch-D, Tranport and Environment, **Vol. 5, No. 2, pp. 121-135, (2000).**

59. Norros I., A storage model with self similar input, Queuing Systems,**Vol.16,pp 387- 396, (1994).**

60. Norros I., On the use of Fractional Brownion Motion in the theory of Connection less Network, IEEE Journal on selected Areas in Communication, **Vol. 13, pp. 953-962, (1995).**

61. Omekara C. O.*, Okereke O. E., Ukaegeu L. U. Forecasting Liquidity Ratio of Commercial Banks in Nigeria Microeconomics and Macroeconomics 2016, 4(1): **28-36, DOI: 10.5923/j.m2economics.20160401.03.**

62. Paxson, V. and Floyd, S., Wide Area Traffic: The Failure of Poisson Modeling, IEEE/ACM Transactions on Networking, **Vol. 3, pp. 226-244. (1995).**

63. Perati, M.R., Raghavendra, K., Koppula, H.K.R., Doodipala, M.R. and Dasari, R., Self Similar Behavior of Highway Road Traffic and Performance Analysis at Toll Plazas, Journal of Transportation Engineering, **Vol. 138, pp. 1233-1238, (2012).**

64. Pushpalatha Sarla, Mallikarjuna Reddy D., Manohar Dingari, Queue Length-Busy Time Distribution Of Web Users Data With Self Similar Behavior, International Journal of Research in Engineering and Technology, **Vol. 05, No. 05, (2016).**

65. Panarat SRISAENG , Glenn S. BAXTER, Graham WILD, Forecasting demand for Low cost Carriers In Australia Using an artificial Neural Network Approach, AVIATION, 1648-7788/eISSN1822-4180, 201,**Volume19(2):90-103,doi:10.3846/16487788.2015.1054157.**

66. Park H, Forecasting Three-Month Treasury Bills Using ARIMA and GARCH Models, Econ 930, Department of Economics, Kansas State University, 1999.

67. Parrelli R, "Introduction to ARCH & GARCH models", Optional TA Handouts, Econ 472 Department of Economics, University of Illinois, 2001.

68. Qiang Meng and Hooi Ling Khoo, Self-similar characteristics of vehicle arrival pattern on Highways. Journal of Transportation Engineering, **Vol. 135, No. 11, (2009).**

69. Qin Yu, Yuming Mao, Taijun Wang, Fan Wu, Hurst parameter estimation and characteristics analysis of aggregate wireless LAN traffic, Communications Circuits and Systems, , **Vol. 1, pp. 339-345, (2005).**

70. Raj kumar L.P., Sampath Kumar K., Mallikarjuna Reddy D., Malla Reddy Perati, Performance Analysis of Internet router employing partial Buffer sharing Mechanis under

Markovian Modeled Self Similar Variable length packet input traffic, International Journal of Pure and Applied Mathematics, **Vol. 67,No. 4, pp. 407-421,(2011a).**

71. Raj kumar L.P., Sampath Kumar K., Mallikarjuna Reddy D., Malla Reddy Perati, Analytical Model for Loss and Delay Behavior of the Switch under Self Similar Variable Length Packet Input Traffic, International Journal for Computer Science, **Vol. 38, No. 1, pp. 26-31, (2011b).**

72. Robert S. and Le Bodec J.Y., on a Markov modulatedd chain Exhibitting Self-Similar traffic Over Finite Time scale, Performance evaluation, **Vol. 27, pp. 159-173, (1996).**

73. Robert S. and Le Bodec J.Y., New Models for Pseudo Self Similar Traffic, Performance Evaluation , **Vol. 30, No. 2, pp. 57-68, (1997).**

74. Sampath Kumar K., Malla Reddy P., Adilaxmi T., Modelling and Analysis of WDMOPS Employing Tunable Converter Sharing under Self-Similar Variable Length Packet Traffic, International Journal of Information and Communication Technology Research, **pp.813-819, (2012).**

75. Sarker M.M.A., Estimation of the Self-Similarity Parameter in Long Memory Processes, Journal of Mechanical Engineering, **Vol. 38, pp. 32-37, (2007).**

76. Sarla P., Doodipala M. and Dingari, M., Self Similarity Analysis of Web Users Arrival Pattern at Selected Web Centers, American Journal of Computational Mathematics, **Vol. 6, pp. 17-22, (2016).**

77. Shao S.K., Malla Reddy Perati, Tsai M.G., Tsao H.W. and Wu J., Generalized variance-based Markovian fitting for self-similar traffic modeling, IEICE Trans.Commun., **Vol. E88-B, No. 12, pp.659-4663, (2005).**

78. Shibin Song, J.K.Y. Ng J.K.Y., Bihai Tang, Some on the self-similarity property in communication networks, Communications IEEE Transactions, **vol. 52, No. 10, pp. 1636-1642, (2004).**

79. Shou-Kuo Shao, Malla Reddy Perati, Meng-Guang Tsai, Hen-Wai Tsao and Jingshown Wu, Generalized Variance-Based Fitting for Markovian Type Self Similar Traffic Modeling, IEICE Trans.Commun, Special Issue on Internet Technology-V, **Vol. E88-B, No. 4,pp. 1493- 1502, (2005).**

80. Shuguo Yang, Xiaoyan Yang, The Application of the Queuing Theory in the Traffic Flow of Intersection, International Journal of Mathematical and Computational Sciences, **Vol. 8, No.6, pp. 984-989, (2014).**

81. Souza, G. P. Samohyl, R. W. Miranda, R. G.,Métodossimplificados de previsãoEmpresarial Rio de Janeiro, RJ: Ciência ModernaLtda, **2008, 181pp.**

82. Tong H., "Threshold Models in Non-Linear Time Series Analysis", Springer-Verlag, New York, 1983.

83. Tsybakov, B., Georganas N.D., Self-similar processes in communications networks, Information Theory IEEE Transactions, **Vol. 44, No. 5, pp. 1713-1725, (1998).**

84. Van Woensel, T., Kerbache, L., Peremans, H. & Vandaele, N., Vechicle routing routing with dynamic travel times : A queuing approach, European Journal of Operational Research, Elsevier, **Vol. 186, No. 3, pp. 990-1007, (2008).**

85. Vern Paxson, Sally Floyd, Wide Area Traffic: The Failure of Poisson Modeling, IEEE/ACM Trans. Networking, **Vol. 3, pp. 226-244, (1995).**

86. Willinger W., Taqqu M.S., Sherman R.and Wilson D.V., Self Similarity through High variability: Statistical Analysis of Ethernet LAN Traffic at the source level, IEEE/ACM Trans. Networking, **Vol. 5, No. 1, pp.71-86, (1997).**

87. Wolfang Fischer, Kathlean Meier-Hellstern,The Markov Modulated Poisson Process (MMPP) Cookbook, Performance Evaluvation, **Vol. 18, No. 2, pp. 149-171, (1992).**

88. Willinger W., Taqqu M.S., Sherman R.and Wilson D.V., Self Similarity through High variability: Statistical Analysis of Ethernet LAN Traffic at the sourc level, IEEEE/ACM Trans. Networking, **Vol.5, No.1, pp.71-86, Jan (1997).**

89. Xiaolong Jin, Geyong Min, Modelling and Analysis of Priority Queuing Systems with Multi-Class Self-Similar Network Traffic: A Novel and Efficient Queue-Decomposition approach, Communications IEEE Transactions, **Vol. 57, No. 5, pp. 1444-1452, (2009).**

90. Xiaolong Jin, Geyong Min, Speros Velentzas, Jianmin Jiang, Quality-of-Service Analysis of Queuing Systems with Long-Range-Dependent Network Traffic and Variable Service Capacity,Wireless Communications IEEE Transactions on, **Vol. 11, No. 2, pp. 562-570, (2012).**

91. Yoshihara T., Kasahara S. and TakahashiY., Practical time-scale fitting of self- similar traffic with Markov-modulated Poisson process, Telecommunication Systems, **vol. 17, No. 1, pp. 185-211, (2001).**

92. Yung-Chang Wang, Chich-Wei Lin, Chung-Chin Lu, Loss Behavior in space Queue with Batch Markovian Arrival Process Discrete Time Case, Performance Evaluation , **Vol. 41, pp. 269-293, (2000).**

93. Zhi Quan, Jong-Moon Chung, A novel analysis of queue length in differentiated services networks with self-similar arrival processes, Circuits and Systems, MWSCAS, **Vol. 3, pp. I-III, (2002).**

94. Zhang G.P, "A neural network ensemble method with jittered training data for time series forecasting", Information Sciences 177 (2007), **pages: 5329–5346.**

95. Zhang G.P, "Time series forecasting using a hybrid ARIMA and neural network model", Neurocomputing 50 (2003), **pages: 159–175.**

B) Thesis

1. Hae-Duck Joshua Jeong, *Modelling of Self-Similar Teletraffic for Simulation*, **Ph.D Thesis, University of Canterbury, New Zealand, (2002).**
2. LP. Raj kumar ,Analytical Model for Performance Study of the Switch under Self-Similar Variable Length Packet Traffic **Ph.D Thesis, Kakatiya University , Warangal (2011).**
3. D. Mallikarjuna Reddy, Performance Analysis of Internet Router under Self-Similar Variable Packet Length Input Traffic-Markovian Approach, **Ph.D Thesis, Kakatiya University , Warangal (2012).**

C) Books

1. Anderson, T. W., The Statistical Analysis of Time Series, **Wiley, New York, (1971).**
2. Bahman Zohuri ,Dimensional Analysis and Self Similarity Methods for Engineers and Scientists, **Cham, Switzerland, Springer, (2015).**
3. Beran J., Statistics for Long-Memory Processes, **Chapman and Hall, (1994).**
4. Bhat,U.N., An Introduction to Queuing Theory,**2nd Edition,Springer,India,(2015).**
5. Box G. E. P., and G. M. Jenkins, Time Series Analysis: Forecasting and Control, Revised Edition, **Holden-Day, San Francisco, (1976).**
6. Box G. E. P., and G. M. Jenkins, Time Series Analysis, Forecasting and Control, **Holden-Day, San Francisco, CA, 1970.**
7. Box G. E. P., G. M. Jenkins and D. W. Bacon, Models for Forecasting seasonal and non seasonal time series, in Spectral Analysis of Time Series, ed. B. Harris, **Wiley, New York, (1967).**

8. Brockwell P.J., Davis R.A., Times Series, Theory and Methods, Springer Series in Satistics, **New York, Springer –Verlag, (1987).**
9. Brockwell, P.J. and Davis, R.A., An Introduction to Time Series and Fore-Casting. **Springer - Verlag, New York, (1996).**
10. Dattatreya ,G.R., Performance Analysis of Queuing and Computer Networks, **Chapman and Hall, New York, (2008).**
11. Doob, J. L., Stochastic Processes **Wiley, New York, (1953).**
12. Grenander, U., and M. Rosenblatt, Statistical Analysis of Stationary Time Series, **Wiley, New York, (1957).**
13. James Douglas Hamilton,Time Series Analysis, **Princeton university press, (1994).**
14. Latouche G.,and Ramaswami V., Introduction to matrix Analytic methods in Stochastic modeling , **SIAM Press, Philadelphia, Pennsylvania, (1999).**
15. Mandolbrot B.B., The Fractal Geometry of Nature, **W.H.Freeman (E.d), New York, (1982).**
16. Markidakis, S., Wheelwright S.C. and Hydman, R.J. Forecasting Methods and Applications. **3rd Edition, John Wiley & Sons. Inc., Hoboken, (1998).**
17. Montgomery, D. C., Peck, E. A. and Vining, G., Introduction to Linear regression Analysis. **3rd Edition, New York, John Wiley & Sons, (2001).**
18. Montgomery, D.C., Jennings, C.L. and Kulachi, M., Introduction to Time Series Analysis and Forecasting, **Wiley Blackwell, New Jersey, (2008).**
19. Neuts, Matrix-Geometric Solutions in Stochastic Model, An Algorithemic Approach, **Dover Publications, (1995).**
20. Park k., and Willinger W., eds. Self-similar network traffic and performance evaluation, **Wiley, New York, (2000).**
21. Paul G. Hoel , Sidney C. port and Charles J. Stone, Introduction to Stochastic Processes, **waveland, press, (1986).**
22. Peitgen H.O.,Jurgens,H.,andsaupe,D.Chaos and fractals: New Frontiers of Science **Springer-Verlag, (1992).**
23. Shumway, R.H, stoffer, D.S., Time series Analysis and Its Applications, Springer Texts in Statistics , New York: **Springer –Verlag, (2000).**
24. Venkatarama Krishnan and Kavitha Chandra, Probability and Random Processes, **second edition, John Wiley & Sons, New Jersey, Canada,(2015)**
25. Willinger.W and Kihong Park,Self Similar Network Traffic and Performance Evaluation, **A wiley –Interscience publication, (2000).**

26. Willinger W., Taqqu M.S and Erramilli A., A Bibilographical Guide to Self Similar Traffic and Performance Modeling for Modern High Speed Networks,Stochastic Networks: Theory and Applications, **Oxford University Press, (1996).**

D) **Conference proceedings**

1. Addie R.,Zukerman M., and Neame J., Fractal Traffic Measurement, Modeling and Performance Evaluation, Proceeding of IEEE INFOCOMM, **(1995).**
2. Chih-How Chang, Shou-Kuo Shao, Malla Reddy Perati and Jingshown Wu, Performance Study of Various Packet Scheduling Algorithms for Variable-Packet- Length Feedback Type WDM Optical Packet Switches , Proceeding of IEEE HPSR, **(2006).**
3. Gospodinov, Evgeniya Gospodinova, The graphical methods for estimating Hurst parameter of self Similar netwrok traffic, International Conference on computer systems and Technologies, **(2005).**
4. Jing Xiao, Xiaodong Ji, Xing Zhang, Wenbo Wang, Cross-layer performance analysis in cooperative relay networks with poisson and self-similar traffic input, Wireless Information Technology and Systems (ICWITS) IEEE International Conference, **pp. 1-4, (2010).**
5. Joarder R.A., Analysis of Real-Time Multimedia Traffic in the Context of Self-Similarity, In Proceedings of International Conference on Electrical and Computer Engineering (5th ICECE), **(2008).**
6. Lau,W.C., Erramilli, A., Wang,J.,and Willinger, W., self Similar Traffic Generation: the Random Midpoint Displacement Alogoritham and Its Properties. I*n Proceedings of IEEE ICC'95*, **pp. 466-472, (1995).**
7. Likhanov N., Tsybakov B. and Georganas N.D., Analysis of an ATM buffer with self-similar ("fractal") input traffic, proceedings of INFOCOM'95, **(1995).**
8. MallaReddy Perati, Chih-How Chang, Shou-Kuo Shao and Jingshown Wu,An Efficient Approximate Markovian Model for Optical Packet Switched Employing Partial Buffer Sharing Mechanism under Self similar Traffic Input, *Proceedings of IEEE HPSR*, **(2007).**
9. Mitko Gospodinov, Evgeniya Gospodinova,The graphical methods for estimating Hurst parameter of self-similar network traffic, *International Conference on Computer Systems and Technologies,* **(2005).**

10. Muhammad Marsudi , Noorazam bin Azman, Queuing theory application to analyze production capacity, *Second International Conference on Engineering System Management and Applications, IEEE,* **pp. 1-6 (2010).**

11. Pushpalatha Sarla, Mallikarjuna Reddy,D., Manohar Dingari, A Study on Self Similarity Analysis of Web Users Data at Selected Web Centers, *Proceedings of International conference on Mathematics,* **(2015).**

12. Song S., Ng J.K.Y., Bihai Tang, On the self-similarity property of the output process from a network server with self-similar input traffic, *Sixth International Conference on Real-Time Computing Systems and Applications,* **pp. 128-132, (1999).**

13. Xiaolong Li, Hancheng Lu, Hao Lu, QoS Analysis of Self-Similar Multimedia Traffic with Variable Packet Size in Wireless Networks, Vehicular Technology Conference (VTC Fall), IEEE 78th, **pp 1-5, (2013).**

E) e-Websites/downloads

1. http://www.dgca.nic.in
2. https: //www.newgenapps.com
3. http://en.wikipedia.org/wiki/self similarity.
4. http://ntptel.ac.in/courses
5. www.statistics solutions.com/what is –linear regression/
6. https: //www.science direct.com.
7. https://en.wikipedia.org/wiki/Autoregressive_ fractionally_integrated_ moving_average.
8. https://otexts.com/fpp2/holt-winters.html.
9. https://www.investopedia.com/terms/n/neuralnetwork.asp.
10. https://en.wikipedia.org/wiki/Air_India
11. https://en.wikipedia.org/wiki/Indigo.

APPENDICES

Appendix A

1. Math Lab execution code linked with the periodogram method:

```
function H = per(sequence,isplot)
%
% 'per' estimate the hurst parameter of a given sequence with
periodogram
%     method.
%
% Inputs:
%     sequence: the input sequence for estimate
%     isplot: whether display the plot. without a plot if isplot
equal to 0
% Outputs:
%     H: the estimated hurst coeffeient of the input sequence


if nargin == 1
    isplot = 0;
end

n = length(sequence);
Xk = fft(sequence);
P_origin = abs(Xk).^2/(2*pi*n);
P = P_origin(1:floor(n/2)+1);

x = log10((pi/n)*[2:floor(0.5*n)]);
y = log10(P(2:floor(0.5*n)));

% Use the lowest 20% part of periodogram to estimate the similarity.
X = x(1:floor(length(x)/5));
Y = y(1:floor(length(y)/5));
p1 = polyfit(X,Y,1);
Yfit = polyval(p1,X);
H = (1-(Yfit(end)-Yfit(1))/(X(end)-X(1)))/2;

if isplot ~= 0
    figure,clf,hold on;
    plot(x,y,'b.');
    plot(X,Yfit,'r-','LineWidth',3);

xlabel('log10(Frequency)'),ylabel('log10(Periodogram)'),title('Perio
dogram Method');
end
```

2. Math Lab execution code linked with the residuals of regnression method:

```
function H = peng(sequence,isplot)
%
% 'peng' estimate the hurst parameter of a given sequence with
residuals
%     of regression method.
%
% Inputs:
%     sequence: the input sequence for estimate
%     isplot: whether display the plot. without a plot if isplot
equal to 0
% Outputs:
%     H: the estimated hurst coeffeient of the input sequence

if nargin == 1
    isplot = 0;
end

% Calculate aggregate level.
N = length(sequence);
FBM = cumsum(sequence);
mlarge = floor(N/5);
msmall = max(10,log10(N)^2);
M = floor(logspace(log10(msmall),log10(mlarge),50));
M = unique(M);
n = length(M);
cut_min = ceil(n/10);
cut_max = floor(7*n/10);

% Calculate residuals under different aggregate level.
Goble_residuals = zeros(1,n);
for i = 1:n
    m = M(i);
    k = floor(N/m);
    matrix_FBM = reshape(FBM(1:m*k),m,k);
    x = 1:m;
    Local_residual = zeros(1,k);

for j = 1:k
        y = matrix_FBM(:,j);
        vv=[x' ones(length(x),1)];
        p=vv\y;
        norm_xx=norm(y-vv*p);
        Local_residual(j) = norm_xx.^2/m;
end

     Goble_residuals(i) = mean(Local_residual);
end


% Fit and calculate H.
x = log10(M);
y = log10(Goble_residuals);
X = x(cut_min:cut_max);
Y = y(cut_min:cut_max);

p1 = polyfit(X,Y,1);
Yfit = polyval(p1,X);
yfit = polyval(p1,x);
```

```
H = 0.5*(Yfit(end)-Yfit(1))/(X(end)-X(1));

if isplot ~= 0
    figure,hold on;
    plot(x,y,'b*');
    plot(X,Yfit,'r-','LineWidth',2);
    plot(x(1:cut_min),yfit(1:cut_min),'r:','LineWidth',2);
    plot(x(cut_max:end),yfit(cut_max:end),'r:','LineWidth',2);
    xlabel('Log of Aggregate Level'),ylabel('Log of Residual
Vaiance'),title('Peng Method');
end
```

3. Math Lab execution code linked with the Percentile method:

```
function H = percentile(userspermin,isplot)
if nargin == 1
    isplot = 0;
end
pps_fname = 'userspermin.txt';
[sequence] = textread(pps_fname, '%u');
N = length(sequence);
for i=1:100
    p(i)=i*N/100+0.5;
    disp(p(i));
    i=i+1;
end
```

Appendix B

The number of Passengers travelled monthly by AIR INDIA Scheduled Domestic Flights during January 2012 to December 2018.

Sno.	Month/Year	passengers	Sno.	Month/Year	passengers	Sno.	Month/Year	Passengers
1	**Jan-12**	796467	29	**May-14**	983898	57	**Sep-16**	1117959
2	**Feb-12**	738710	30	**Jun-14**	931132	58	**Oct-16**	1030490
3	**Mar-12**	767293	31	**Jul-14**	885447	59	**Nov-16**	1068768
4	**Apr-12**	773982	32	**Aug-14**	878178	60	**Dec-16**	1245063
5	**May-12**	756425	33	**Sep-14**	909531	61	**Jan-17**	1265046
6	**Jun-12**	754231	34	**Oct-14**	1079368	62	**Feb-17**	1084609
7	**Aug-12**	684959	35	**Nov-14**	959403	63	**Mar-17**	1089418
8	**Jul-12**	691686	36	**Dec-14**	1133456	64	**Apr-17**	1086116
9	**Sep-12**	669342	37	**Jan-15**	1104361	65	**May-17**	1219997
10	**Oct-12**	852892	38	**Feb-15**	1010005	66	**Jun-17**	1149333
11	**Nov-12**	913426	39	**Mar-15**	996246	67	**Jul-17**	1179933
12	**Dec-12**	880300	40	**Apr-15**	999526	68	**Aug-17**	1176397
13	**Jan-13**	907261	41	**May-15**	1043708	69	**Sep-17**	1163026
14	**Feb-13**	804577	42	**Jun-15**	934155	70	**Oct-17**	1216153
15	**Mar-13**	907432	43	**Jul-15**	1032276	71	**Nov-17**	1271612
16	**Apr-13**	852743	44	**Aug-15**	1058866	72	**Dec-17**	1298442
17	**May-13**	933573	45	**Sep-15**	990986	73	**Jan-18**	1369342
18	**Jun-13**	809681	46	**Oct-15**	1015358	74	**Feb-18**	1259379
19	**Jul-13**	803943	47	**Nov-15**	1119595	75	**Mar-18**	1365653
20	**Aug-13**	908224	48	**Dec-15**	1217671	76	**Apr-18**	1345592
21	**Sep-13**	809330	49	**Jan-16**	1157645	77	**May-18**	1333851
22	**Oct-13**	799041	50	**Feb-16**	1085634	78	**Jun-18**	1238131
23	**Nov-13**	861128	51	**Mar-16**	1086743	79	**Jul-18**	1255241
24	**Dec-13**	911763	52	**Apr-16**	1117795	80	**Aug-18**	1265214
25	**Jan-14**	874549	53	**May-16**	1246137	81	**Sep-18**	1173238
26	**Feb-14**	839258	54	**Jun-16**	1156490	82	**Oct-18**	1276807
27	**Mar-14**	951993	55	**Jul-16**	1177504	83	**Nov-18**	1242223
28	**Apr-14**	856146	56	**Aug-16**	1134913	84	**Dec-18**	1372904

(Source: Official Website of DGCA, http://www.dgca.nic.in).

The number of Passengers travelled monthly by INDIGO Scheduled Domestic Flights during January 2012 to December 2018.

Sno.	Month/Year	passengers	Sno.	Month/Year	passengers	Sno.	Month/Year	Passengers
1	**Jan-12**	1111061	29	**May-14**	1908334	57	**Sep-16**	3295826
2	**Feb-12**	1076699	30	**Jun-14**	1797101	58	**Oct-16**	3692828
3	**Mar-12**	1068637	31	**Jul-14**	1599977	59	**Nov-16**	3772583
4	**Apr-12**	1211875	32	**Aug-14**	1858664	60	**Dec-16**	3848322
5	**May-12**	1356436	33	**Sep-14**	1907378	61	**Jan-17**	3809228
6	**Jun-12**	1328497	34	**Oct-14**	1890273	62	**Feb-17**	3418605
7	**Aug-12**	1225859	35	**Nov-14**	1967992	63	**Mar-17**	3611371
8	**Jul-12**	1205383	36	**Dec-14**	2324221	64	**Apr-17**	3778780
9	**Sep-12**	1093688	37	**Jan-15**	2276404	65	**May-17**	4190914
10	**Oct-12**	1266126	38	**Feb-15**	2230645	66	**Jun-17**	3825814
11	**Nov-12**	1368563	39	**Mar-15**	2286128	67	**Jul-17**	3699451
12	**Dec-12**	1456630	40	**Apr-15**	2481285	68	**Aug-17**	3678245
13	**Jan-13**	1408012	41	**May-15**	2769283	69	**Sep-17**	3664509
14	**Feb-13**	1341210	42	**Jun-15**	2536554	70	**Oct-17**	4133027
15	**Mar-13**	1423569	43	**Jul-15**	2416916	71	**Nov-17**	4131844
16	**Apr-13**	1511094	44	**Aug-15**	2384943	72	**Dec-17**	4430070
17	**May-13**	1685168	45	**Sep-15**	2430449	73	**Jan-18**	4556904
18	**Jun-13**	1480879	46	**Oct-15**	2589861	74	**Feb-18**	4290189
19	**Jul-13**	1445248	47	**Nov-15**	2597765	75	**Mar-18**	4576236
20	**Aug-13**	1531406	48	**Dec-15**	2743325	76	**Apr-18**	4579916
21	**Sep-13**	1378691	49	**Jan-16**	2725711	77	**May-18**	4852909
22	**Oct-13**	1510184	50	**Feb-16**	2754131	78	**Jun-18**	4672686
23	**Nov-13**	1467763	51	**Mar-16**	3023228	79	**Jul-18**	5002416
24	**Dec-13**	1575872	52	**Apr-16**	3066556	80	**Aug-18**	4757378
25	**Jan-14**	1426580	53	**May-16**	3336839	81	**Sep-18**	4920335
26	**Feb-14**	1464070	54	**Jun-16**	3023081	82	**Oct-18**	5074853
27	**Mar-14**	1601141	55	**Jul-16**	3383768	83	**Nov-18**	5005919
28	**Apr-14**	1679963	56	**Aug-16**	3341081	84	**Dec-18**	5478523

(Source: Official Website of DGCA, http://www.dgca.nic.in).

List of Publications & Conference Proceedings

1. Manohar Dingari, D. Mallikarjuna Reddy, V. Sumalatha, " Air Traffic Forecasting Using Time Series Models" *International Journal of Recent Technology and Engineering* **(ISSN: 2277-3878), Vol. 8, Issue.4, November 2019 page No: 1061-1065.**

2. Manohar Dingari, D. Mallikarjuna Reddy, V. Sumalatha, "Time series Analysis for Long Memory Process of Air traffic Using ARFIMA" *International Journal of Scientific and Technology Research* **(ISSN: 2277-8616), Vol. 8, Issue.10, October 2019 page No: 395-400.**

3. Manohar Dingari, D. Mallikarjuna Reddy, V. Sumalatha, "Air traffic Forecfasting Using Artificial Neural Networks" *International Journal of Scientific and Technology Research* **(ISSN: 2277-8616), Vol. 8, Issue.10, October 2019 page No: 556-559.**

4. Manohar Dingari, D. Mallikarjuna Reddy, V. Sumalatha, "Self-Similar Performance of Passengers Arrival Patterns of Air India-Domestic Services" *International Journal of Research And Analytical Reviews* **(E-ISSN: 2348-1269), Vol. 6, Issue.2, June 2019 page No: 191-195.**

5. V. Sumalatha, Manohar Dingari, C. Jayalakshmi, "Short Term Road Traffic Flow Forecfasting Using Artificial Neural Network" *International Journal of Recent Technology and Engineering* **(ISSN: 2277-3878), Vol. 8, Issue.3, September 2019 page No: 7998-8000 .**

6. Pushpalatha Sarla, D. Mallikarjuna Reddy, Manohar Dingari,"Self- similar Behaviour Highway Traffic Analysis –Using Queuing Systems" *International Journal of Computer Sciences and Engineering* **(ISSN: 2347-2693), Vol. 7, Issue.4, April 2019.**

7. Pushpalatha Sarla, D. Mallikarjuna Reddy, Manohar Dingari, "Self Similarity Analysis of Web Users Arrival Pattern at Selected Web Centers". *American Journal of Computational Mathematics,* **Vol. 6, No.1, Mar 2016.**

8. Pushpalatha Sarla, D. Mallikarjuna Reddy, Manohar Dingari, "Queue Length-Busy Time Distribution of Web Users Data With Self Similar Bihavior" **Vol 19, Issue 5, Ver. 5, pp 84-89, 2016** *IJRET: International Journal of Research in Engineering and Technology.*

9. Manohar Dingari, D. Mallikarjuna Reddy, V. Sumalatha, "Analysis of Air Traffic Data Using Time Series Models" *Proceedings of the International Conference of Mathematical Sciences and Applications* -ICMSA 2019, **9th-11th August 2019**, GITAM University, Hyderabad.

10. Manohar Dingari, D. Mallikarjuna Reddy, V. Sumalatha, "Self-Similar Behavior of Passengers Arrival Patterns of Air India-Domestic Services" *Proceedings of the International Conference of Mathematical Sciences and Applications* - ICMSA 2019, **9th-11th August 2019**, GITAM University, Hyderabad.

11. Pushpalatha Sarla, D. Mallikarjuna Reddy, Manohar Dingari, "Queue Length-Busy Time Distribution of Web Users Data with Self Similar Behavior". *Proceedings of the International Conference on Innovations and Advancements in Computing* -ICIAC 2016, **18th-19th March 2016**, GITAM University, Hyderabad.

12. Pushpalatha Sarla, D. Mallikarjuna Reddy, Manohar Dingari,"Self Similarity Analysis of Web Users Arrival Pattern at Selected Web Centers". *International Conference on Mathematics,* **26th-28th November, 2015,** University of Kerala, Kerala

Kolibri 360 GmbH
Ferdinand-Jühlke-Straße 7
99095 Erfurt, Deutschland
produktsicherheit@kolibri360.de

Druck:
CPI Druckdienstleistungen GmbH
im Auftrag der
Kolibri 360 GmbH
Ferdinand-Jühlke-Str. 7
99095 Erfurt